"Stevie Famulari, Gds's *Equity and Inclusion in Green Design: Inclusive Approaches for Designing Healthier Work and Living Spaces* is a vital resource for anyone committed to creating equitable and sustainable spaces. She beautifully addresses the critical intersection of green design and social inclusivity, moving beyond superficial aesthetics to delve into the needs of the most impacted communities. Stevie delivers a transformative vision, empowering designers, advocates, and policymakers to craft spaces where everyone thrives. This book is not simply a guide; it is a push to lead differently and to build with and for the community. Whether you're a skilled professional or just beginning your journey, Stevie's insights and practical framework will propel you toward creating the equitable spaces that we all deserve."

Janna Pea
Entrepreneur + Narrative Change Expert
PeaNation.co

"Diving deeper into green design, Stevie Famulari, Gds explores green design solutions that address equity and inclusion. Her fourth book continues an amazing journey using her insightful knowledge and voice to illustrate how green design can transform exterior and interior public and private spaces to be inclusive for people. Throughout multiple cultures, living design can bring people together and celebrate cultures to grow and thrive. Stevie Famulari, Gds illustrates how green design has failed in the past and present and, more importantly, how to use it to change the present and future. Individuals, businesses, administrators, and communities can apply the healthy green practices in this book to find solutions that offer representation, solving for the respective needs of groups and institutions looking to transcend the hurdles."

Henry A. Obispo
Eco-Gastronomer
Founder of BIOME & ReBORN FARMS

"In *Equity and Inclusion in Green Design: Inclusive Approaches for Designing Healthier Work and Living Spaces*, Stevie Famulari, Gds continues to expand on what green design is and can be. In her latest book in the green design collection, Famulari explores the important topic of equity and inclusion in healthy design, illustrating some of these vital issues and also how change can be addressed through green design. Famulari's approach is relatable and practical, from small private sites to large public spaces. As a professor who teaches landscape architecture, urban design, and horticulture, I see this book as an essential tool for students to learn inclusive design practices from the beginning of their careers as well as for non-designers, administrators, agencies, practitioners, and everyone who wants to use design as an act of change for equity and inclusion."

Hye Yeon Park, PhD, Professor
Mississippi State University, LAED Dept

Equity and Inclusion in Green Design

This book focuses on seeing how inequity and lack of inclusion presently shows in existing spaces, how to retrofit existing spaces to correct inequity and lack of inclusion through green design, as well as design new green spaces with a focus of equity and inclusion. This includes addressing equity and inclusion focusing on people's health, abilities, sexuality, gender, age, race, language, income, and culture which show in public and private site design, more so in the lack of equality and inclusion in green design.

The effects of equity and inclusion show in multiple ways, including the lack of:

- Ease of access to affordable healthy food and agriculture.
- Community green design for people with different health and abilities.
- Ease of access in public and private spaces for people with different visual, walking, hearing, mobile, and learning abilities.
- Safe and equitable public green design for womxn and people in the LGBTQIA+ community.
- Green design in areas with a focus for people who are Black, Brown, and from underserved communities.
- Ease of access in public and private spaces for people who do not fluently understand English or the native language of the site.
- Green spaces in areas of people with lower economic means.
- Community green design for immigrants to celebrate their culture.

Addressing equity and inclusion can be designed directly into sites as integral in the design process through retrofitting existing sites, as well as new site design. Exploring healthy approaches to the important topic of equality is an important step in acknowledging the lack of equity and inclusion and exploring new designs. This involves listening to people; and creatively exploring design for diversity and inclusion for health, abilities, sexuality, gender, age, race, language, income range, and cultures.

This book gives readers a way to not only understand equity and inclusion in greening but also to understand how to develop inclusive approaches for design. The two basic site types in this book are: existing spaces, which greening design is applied to afterwards; and upcoming spaces, which greening design is integral to the built into the space. Retrofitting greening into existing space can also be combined with new designs. There are examples of both types throughout the book.

Equity and Inclusion in Green Design

Inclusive Approaches for Designing Healthier Work and Living Spaces

Stevie Famulari, Gds

Routledge
Taylor & Francis Group
A PRODUCTIVITY PRESS BOOK

Designed cover image: Shutterstock

First published 2026
by Routledge
605 Third Avenue, New York, NY 10158

and by Routledge
4 Park Square, Milton Park, Abingdon, Oxon, OX14 4RN

Routledge is an imprint of the Taylor & Francis Group, an informa business

ISBN: 978-1-032-75707-0 (hbk)
ISBN: 978-1-032-75706-3 (pbk)
ISBN: 978-1-003-47529-3 (ebk)

DOI: 10.4324/9781003475293

Typeset in Garamond
by Apex CoVantage, LLC

To the strong, creative, and powerful people
who apply these equitable and inclusive green practices out loud
and **CHANGE THE WORLD**

Contents

Foreword

Ashley Zidon, AICP
Outdoor community developer

I daydream of a future when people feel safe walking home in the dark, everyone can access a park, caretakers can comfortably push a stroller or wheelchair, trees in our neighborhoods lower the extreme heat, people have access to fresh food . . . then I am jolted awake. As someone who has studied cities for nearly 15 years, it is hard to find a place, neighborhood, or community that accomplishes all these things simultaneously.

This book is a call for action. In its simplest form, it is about elevating the voices of people who have been traditionally underrepresented and deploying a framework that will create inclusive places and policies that will reach across communities. *Equality and Inclusion in Green Design: Inclusive Approaches for Designing Healthier Work and Living Spaces* centers the importance of shifting power back to people and empowering individuals to have ownership in design – to have their stake in space. And that practice is more than door knocking and community meetings. It is an intentional shift in practice to elevate community voices, learn from their lived experiences, and understand how built and natural spaces affect the way they live. Seeing this work through the lens of people's health, abilities, sexuality, genders, age, race, language, income, and cultures will help designers, advocates, administrators, policy makers, and students create equitable places and communities. As an added bonus, embedded in this framework is a foundation of health for both people and the environs, a specialty of Famulari.

As designers, artists, landscape architects, planners, architects, community builders, change leaders, and agitators, we are enormously privileged to work alongside communities: listening, learning, understanding what their communities need and how we can help people thrive. Yet, not all feel this calling to uplift community voice to lead change. Some step in, without a nod to the community, and perpetuate inequities and create spaces for the few. If you feel as if you have intentionally or unintentionally been part of such harm, I invite you to reflect on the role you played and encourage you to keep reading the following chapters.

Stevie Famulari, Gds is a light. Stevie is bold, brilliant, courageous, and a compassionate individual. Stevie has taught in educational settings for decades and is a memorable professor who makes a lasting impact in the classroom and studio by the way she delivers curriculum, design critiques, and constructive dialogue on social inequities. I was one of her students.

My conversations with Stevie are often about projects and design, but are also meaningful exchanges and a place of support. There are some people you are lucky enough to meet in your lifetime, those who lift you up, challenge you, cry with you, sit with you in the silence. Stevie is one of them, and this book is a reflection of support, but this time for a greater community.

Famulari's contribution to the field is rich and transformative. As an author of three previous books on greening, she is an artist, researcher, professor, and founder and principal of her own practice. Stevie Famulari, Gds thoughtfully introduces a vision to easily integrate equity and greening into practice in this new book.

It is incredibly gratifying to watch change happen, to empower people to use their voice for impact, to build and construct places with and for community members, and to better the lives of those folks who frequent and visit these places – and yet, relinquishing that control and transferring that power is not easy. Part of your own privilege has led you here. It is my belief that you are seeking to learn more or beginning your learning journey on how to create meaningful places where all people can thrive. This book is your permission to lead differently. If you are unsure how, the following chapters will offer a new easy-to-follow framework on how to approach your work in creating equitable places. It is my belief that this book will help us craft that idyllic community in my daydream.

Famulari delivers one of her most important lectures yet. Keep this book close and let it propel you to a new chapter of your practice. Or if you are not ready, step aside and gift it to a change-maker who will – because people and our communities depend on it.

About the Author

Stevie Famulari, Gds is an artist, author, researcher, green design specialist, founder and principal of *Engaging Green*, and professor of landscape architecture, horticulture, urban design, remediation, and public art. She has been a professor at the University of New Mexico (UNM), North Dakota State University (NDSU), SUNY Farmingdale, and the University of Cincinnati. She guest lectures at Harvard University, Northeastern University, University of Minnesota, University of Cincinnati, and UC Berkeley. She is a keynote speaker, receives numerous awards, and shares information to help numerous people experience greening in new and unique ways.

This is Stevie's fourth book. Her first book, *Green Up! Sustainable Design Solutions for Healthier Work and Living Environments*, explores unique greening solutions and practices that help create a lifestyle shift, improving the health of living and working spaces for their occupants from individual, business, ecological, and profitable perspectives. Her second book, *Designing Green Spaces for Health: Using Plants to Reduce the Spread of Airborne Viruses*, focuses on using plants in spatial design to reduce the infectiousness of viruses in different working and living spaces. It presents strategies for interior and exterior green designs with plants that are effective for flu virus tolerance and reduction of infectiousness. Her third book, *Ways of Greening: Using Plants and Gardens for Healthy Work and Living Surroundings*, explores designs which rethink work and living spaces with healthy green designs focusing on adding 10% of living green design per square foot to interior and exterior sites. Through the illustrations, designs, photographs, and text, readers can understand and apply greening to their spaces.

Her installation *Engaging Urban Greening* was installed at the Smithsonian Museum in Washington, DC. More than 1,400 seeded paper pieces (colorful papers embedded with a wildflower seed mix) were given to the public during the installation. People from the District of Columbia, New York, Michigan, Maryland, Florida, California, Oregon, Nigeria, Australia, India,

England, and South Korea added green to their community by planting their seeded papers in the soil, creating their own living artwork.

Famulari's designs and design research explore the relationship, extension, and application of green designs and public art to other diverse fields to create healthy spaces for living and working. Time, change, and engaging in the process are some of the continuing themes in her works. Defying traditions and celebrating ecological processes, and using community and individual stories of sites are themes she is widely recognized for. With the applied art of green design combined with the science of phytoremediation, her works have aesthetic beauty, as well as healing properties for both people and the environs.

Her work in greening designs, research, and education can be seen nationally and internationally and includes green walls, planted roofs, green office designs, and green remediation designs for interior and exterior applications; designs for the U.S. Environmental Protection Agency (EPA) in Colorado and Ghost Ranch Visitor Center for the Georgia O'Keefe Museum in Abiquiú, New Mexico; developing green design and policies on the UNM campus; and flood control design for the Red River in North Dakota.

Lectures, presentations, and exhibitions of her work include Plains Art Museum, San Diego Museum, Minnesota Erosion Control Association (MECA), Engineering Mechanics Institute, University of Minnesota (UMN), NDSU, American Society of Landscape Architecture (ASLA), the American Institute of Architecture (AIA), and numerous galleries. Awards of Stevie Famulari, Gds's works and community leadership include the Environs Leadership Award from MECA, Community Service Award from The Arts Partnership (TAP), selected for team USA for the Festival Du Voyageur, Art in Public Places in New Mexico, New York State Floral Industry (NYSFI), Long Island Nursery and Landscape Association (LINLA), and NMEH Award for New Mexico Symphony Orchestra.

Stevie Famulari, Gds's phytoremediation database of hundreds of plants that clean the air, soil, and water of contaminants is used by the EPA; courses at Harvard Graduate School of Design; International Phytoremediation Society; landscape architecture and engineering firms; and administrative agencies. Based on her second book, Stevie also created a database of plants for reducing airborne viruses. She has been an investigator for grants that explore water remediation design for oil drilling and fracking processes, improving air quality, remediation design for communities, and interior greening applications.

Her designs and design research explore the relationship, extension, and application of green designs to other diverse fields. She received her graduate degree in landscape architecture (MLA) from the State University of New York (SUNY) ESF with a concentration in fine arts from Syracuse University. Her undergraduate degree is in fine arts (BFA) from NYU.

Stevie Famulari, Gds's work has appeared in hundreds of books, magazines, newspapers, and television programs, including *Smithsonian Magazine*, Food Network Challenges and Specials, *Oakland Tribune*, World Entertainment News Network, *The Oprah Winfrey Show*, Travel Channel, *Good Morning America*, *CBS Early Morning News*, *Washington Post*, *The Post Standard*, Trust for Public Land, *Boston Herald*, *Berkeley Daily Planet*, *Santa Fe Reporter*, and Minneapolis *Star Tribune*.

Giving Gratitude and Thanks To

For all the people whom I am honored to have as my friends, family, and colleagues, including

Abbott – For being you and leading by example of how to be high road. I do not know where I would be without some of our unique conversations.

Amy and Jason – Thanks for making it easier, all right, all right, all right.

Ashlee – Goddess, for everything I am.

Ashley – For countless dialogues over countless years.

Ayesha – I appreciate the fantastic semester of dialogue and expanding your mind. Keep expanding.

Charlie – I love that you send me reminders – when I've asked for them and when I haven't.

Colin – For saying, "Yes. And how can I help you be *even more* you?" then laughing.

Desi – For twistedly laughing with me in conversations about some topics.

Henry – I appreciate the deep dialogues and the plant that grew with me while I was away.

Hye – For sharing adventures and laughing waaaayyy loudly in our office.

Janna – I love our Sunday brunches no matter where we are.

Kimba – I see you living by example to not judge others and instead choose who you want to be around.

Kris – For growing out loud.

Lisa – For twisted perfect timing and reminders when I need them.

Missy – Goddess, for everything I am.

Newcov Fitness crew, especially Adam, Diane, Emily, Megan – Thanks for a safe and fun space to release things that no longer serve my greatest good.

Nonni – I love you for staying around unconditionally.

Paul and Shane – For dialogues through all our changes and travels.

Rose – For all those weeks of healing and light.

Rusty – For living your dreams to life, reminding me how easy it can be when I let it.

Shelly – Goddess, for everything I am.

Sophie – For reminding me to choose happiness, always.

Travis – For letting me do something good, even *great*, with the process we proudly choose.

UC students, especially Gabby, Guy, James, Rachel, Roozbeh, and Yuni – Your projects and essays will remain in my head for a long, long time. Keep living with infinite ideas, and you will go far.

Viscosity – For the twistedly fun dialogues.

Wally – For making things easier, a green hallway, and for laughing out loud.

Yurika – For making the semester easier. Keep following the things that matter to you.

Introduction

0.1 Equity and Inclusion in Green Design – Acknowledging and Addressing the Issues

Image i.1 **The exteriors of industrialized buildings are opportunities for communities to create a green space that improves the air quality, which the industry may have damaged in its production. This green site can be used to celebrate the arts while using green design for community health.**

This fourth book of the green design book series focuses on seeing how equity and inclusion, or lack thereof, presently show in existing spaces, as well as how to retrofit existing spaces and create new spaces to improve equity and inclusion through green design.

This series of four green design books illustrates different points, perspectives, and directions that green design plays in one's life. Each book stands alone, or they can be used together for understanding and applying green design for a variety of scales, uses, sites, and needs.

DOI: 10.4324/9781003475293-1

This fourth book, *Equality and Inclusion in Green Design: Inclusive Approaches for Designing Healthier Work and Living Spaces*, is created during a loud time in my life when inequity and lack of inclusion in design addressing people's health, abilities, sexuality, gender, age, race, language, income, and culture are so loud that at times it overwhelms me. Inequity and lack of inclusion show in a myriad of fields both in subtle and loud ways, as well as in small words and large events.

Addressing equity and inclusion through green design is where this book intersects with life. While writing and illustrating this book, I believe and trust that things will not just be "better," as better is not equal nor equitable. Better can be anywhere such as barely or partially improving the inequality and lack of equitability. Equal is equal. Nothing short of equality and equity is acceptable. Ever.

Chapter 1 explores ideas and practices addressing health, abilities, sexuality, gender, age, race, language, income, and culture, and how they show in spaces, as well as ideas for change. Terminologies for understanding the book are also included and can be referred to at any point for clarity.

Chapters 2–6 illustrate design approaches to address diversity, equity, inclusion, and justice in designs for people who live daily with inequitable experiences based on their health, ability, sexuality, gender, age, race, language, income, and culture. Case studies of spaces and events which address equity and inclusion, or lack of equity and inclusion in sites and practices are also included in the chapters.

Chapter 7 includes a list illustrating how a lack of equity and inclusion is reflected in spatial design, grouped into the nine categories addressed for equity and inclusion. A second list begins to shed light on design solutions toward diversity, equity, inclusion, and justice across topics such as health, ability, sexuality, gender, age, race, language, income, and culture. Additionally, the chapter provides information on materials and illustrates some design approaches.

This fourth book of the green design book series focuses on seeing how equity and inclusion, or lack thereof, presently show in existing spaces, as well as how to retrofit existing spaces and create new spaces to improve equity and inclusion through green design. Equity and inclusion for health, ability, sexuality, gender, age, race, language, income, and culture can be experienced in public and private site design, more so in their lack of equity in green design. The effects of equity and inclusion show in the following.

- Lack of ease and affordable healthy food, agriculture, and water in underserved populations.
- Lack of green spaces in areas of people with lower economic means.
- Lack of community green design in formerly redlined areas. Redlining is a discriminatory practice in the United States where financial institutions deny funding to families and neighborhoods people that are Black,

Brown, and from underserved communities. This was practiced for years, financially and culturally affecting people not only of earlier generations but also of generations to follow.

- Lack of community green design in areas of populations of immigrants.
- Lack of green design focusing on people who are Black, Brown, and underserved populations.
- Lack of green design affecting health of people with lower incomes.
- Lack of easy access for people to public comforts.
- Lack of access to public recreational design for teenagers and mature populations.
- Lack of ease of access for people who have different abilities for mobility, vision, and reading (both in addressing reading an official country language, as well as people with different reading abilities).

Addressing equity and inclusion can be designed directly into sites as integral in the design process rather than as an afterthought or gesture. This includes both retrofitting existing sites and creating new site designs. Exploring healthy approaches to the important topic of equity is an important step in acknowledging the lack of equity and inclusion, and exploring designs for improved inclusion. The processes involve listening to people and creatively exploring design for diversity and inclusion for addressing people's health, abilities, sexuality, gender, age, race, language, income, and culture.

This book gives readers a way to understand not only equity and inclusion in greening, but also how to develop inclusive approaches to design. The two basic site types in this book are existing spaces to which green design is applied afterwards and upcoming newly developing spaces which greening design can be built into during the process. The first type of retrofitting

Image i.2 **Klapparstigur, Iceland, celebrates the LGBTQIA+ community as inclusive of safe accessible design for the community.**

greening into existing space can also be combined with the second type of space (new designs). There are examples of both types throughout the book.

0.2 The Scope of This Book

This publication continues to build on the knowledge from the previous three green design books, while also being able to stand on its own.

Inspired in part by the articles of multiple months in the Saturday paper, the *New York Times* has explored how equity and inclusion show in different areas, affecting people's health, finances, and culture. The articles include redlined areas, as well as urban areas and non-urban areas. The effects of equity and inclusion are seen through the lack of living green designs in the community affecting people's health and ability to gather in a green community space; lack of access to healthy food and water, also affecting health; and lack of access to waterfronts and other stunning views affecting a sense of belonging to, value in, equality, and inclusion in the community, to name a few ways.

Equity and inclusion are complicated subjects with countless layers to unpack. Specifically, *Equality and Inclusion in Green Design: Inclusive Approaches for Designing Healthier Work and Living Spaces* looks at how equity and inclusion show in existing sites and how green design can help address and improve issues of equity, inclusion, diversity, and justice.

Traditionally and primarily in current practice, green designs are focused on middle-class to upper-class Caucasian cisgender men of good health (Cohen et al., 2021; Nagendra and Sen, 2022). People's health abilities, sexuality, gender, age, race, language, income, and culture – if addressed at all – were afterthoughts or a specialty of very specific site designs. This can be seen in designs throughout centuries and through recent decades.

Image i.3 Weaving small spaces, green roofs, living vertical designs, pocket parks, and large green spaces into a mix of public green spaces for communities to share. These are important for the physical health of people through improved air quality, as well as the emotional health of being grounded in earthy spaces while enjoying the density of urban design.

Image i.4 **Turning a gymnasium into a school and community space for the education of the children and community. This site provides education, food, flowers for sale, and a healthy space for a tired soul to relax and refocus when needed.**

Some examples of the lack of equity and inclusion in green design show in the lack of public green spaces in redlined districts where plants and living materials are disproportionately scarce or entirely missing. Additionally lacking are spaces for people to use for outdoor recreation and meeting other community members. Further, access to agriculture and safe public spaces comforts are lacking for underserved populations, including womxn; children; teenagers mature people; the LGBTQIA+ community (lesbian, gay, bisexual, transgender, queer or questioning their gender, intersex, asexual, and their allies); people who are Black, Brown, and from underserved communities; and people with different abilities.

The effects of the disproportionate amount of green spaces as a lack of equity can be seen in broad strokes, as well as in details in urban design and non-urban design. As a native New Yorker, I can still see the effects of formerly redlined areas in New York where people of lower economic means and people who are Black, Brown, and from underserved populations live. There is a lack of public green space, as well as expanded health issues due to poor air quality because of the lack of green space. Plants help to filter out contaminants in the air. People who live in areas without green spaces disproportionately have health issues when compared to those in spaces of wealth with green spaces (Famulari, 2020; Fernandez, 2006; Nuccitelli, 2023).

0.2.1 Diversity Is a Strength, Inclusivity Is Powerful, Equality Is a Birthright, and Justice Is a Necessity

The green practices in this book focus on the approach of designs that address equity and inclusion in the creation process. This is done by talking with and listening to diverse groups of people, asking questions, and not only listening but truly hearing their wants and needs – and then translating those needs into the spaces where these same people live and work. This process makes an improved and stronger design for a variety of people to enjoy and participate in, and allows the site to change over time.

Some sites will have to be retrofitted with green design with a focus on equity and inclusion, which includes cultural identities; diversity for people with different abilities; and inclusivity for children, teenagers, and mature populations; as well as intentional approaches for those of a range economic means and a range of cultural practices, genders, races, and abilities.

Equity and inclusion of the great range of people and their needs are directions that green design can directly address. Retrofitting existing spaces, as well as the design approach for new spaces, is the approach this author and green design specialist takes throughout this book.

The illustrations of spaces throughout this book are as seen and designed with an approach of equity and inclusion through greening designs. There are urban and non-urban spaces that are illustrated and redesigned through a greening lens. This book opens people's minds to how to see spaces with greening options toward equity and inclusion as unique interiors and exteriors, small and large, horizonal and vertical, and beyond.

References

Cohen, D. A., Williamson, S., & Han, B. (2021). Gender Differences in Physical Activity Associated with Urban Neighborhood Parks: Findings from the National Study of Neighborhood Parks. *Women's Health Issues: Official Publication of the Jacobs Institute of Women's Health*, 31(3), 236–244. https://doi.org/10.1016/j.whi.2020.11.007

Famulari, S. (2020). *Green Up! Sustainable Design Solutions for Healthier Work and Living Environments.* New York: Productivity Press. https://doi.org/10.4324/978042297434

Fernandez, M. (2006, October 29). A Study Links Trucks' Exhaust to Bronx School Children's Asthma. *The New York Times*, Region Section.

Nagendra, H., & Sen, A. (2022). Our Cities are Designed by Men, for Men. *The Third Eye and Science the Wire.* Retrieved 20 February 2025, from https://science.thewire.in/society/gender/our-cities-are-designed-for-men-by-men/

Nuccitelli, D. (2023). The Little-Known Physical and Mental Health Benefits of Urban Trees. *Yale Climate Connections.* Retrieved 12 March 2023, from https://yaleclimateconnections.org/2023/02/the-little-known-physical-and-mental-health-benefits-of-urban-trees/

Chapter 1

Understanding Equity and Inclusion in Green Design

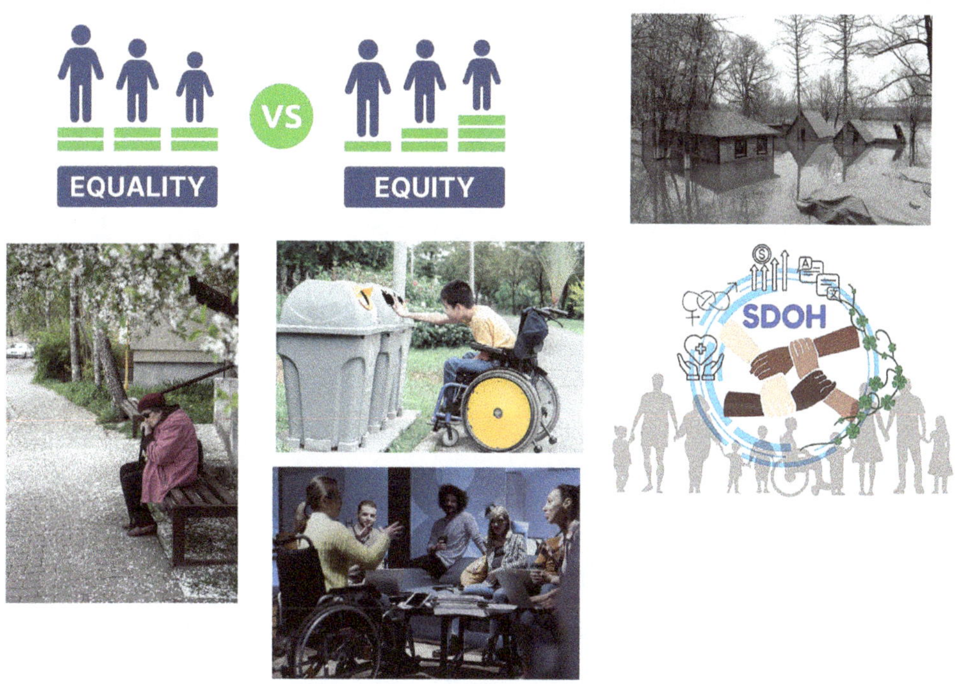

Images 1.1–1.6 **Equity and inclusion as they relate to spatial design include understanding inequity and exclusive spaces, as well as different design solutions to address short-term and long-term change.**

DOI: 10.4324/9781003475293-2

1.1 Introduction

The topics of equity and inclusion in green design are the primary focus of this book. The phrase DEI (diversity, equity, and inclusion) is an acronym that dates to the 1960s in the United States (Kratz, 2024). It was influenced by affirmative action and equality legislation for employees, including Title VII of the Civil Rights Act of 1964, the Equal Pay Act of 1963, and the Discrimination for Employees Act of 1967. The "J" in DEIJ (diversity, equity, inclusion, and justice) was added for social justice much later – for diversity, equity, and inclusion to be effective, social justice is a needed component.

This book does not use the acronyms DEI and DEIJ and instead uses equity and inclusion. This is an intentional choice, as acronyms can be generalized and lose their power. Choosing to use the words equity and inclusion is about taking back the words and their power, definitions, and applications to design. Whether the acronym or grouping of people and ideas have meant well, their inclusion of a wide range of people and ideas can generalize people and ideas and be lost in a larger conglomerate. The point where individuals and ideas are lost in something bigger, rather than empowering something bigger, is where damage happens. Therefore, the terms equity and inclusion are used throughout this book, as well as diversity and justice where applicable.

Further, this book explores green design approaches and green design solutions toward equity and inclusion. It does not address policy or cultural needs in order to address equity and inclusion for the design of spaces. There are subtle and loud policies and practices in which inequity continues in green practices and design of small and large public and private spaces. This first chapter explores how inequity and lack of inclusion presently show in spatial design, and explains ideas and terminology related to inclusion and equity for public and private spaces, as well as green ideas for changing existing spaces.

Health, abilities (different abilities), sexuality, gender, age, race, language, income (low income), and culture are the major components in which there are notable unacceptable differences in equity and inclusion. Green design has the ability to explore, address, change, and improve spaces to directly address some of these inequalities. Further, local, regional, and national policies and practices play an important role that can help to have an effect for justice.

Some issues of health, abilities, sexuality, gender, age, race, language, income, and culture overlap and are combined into layers of challenges in design of public and private spaces. There are also design issues and solutions that interrelate in multiple categories, as well as issues and solutions that are difficult to categorize.

Addressing these topics in green design, public spaces, and private spaces is to benefit and grow culture for inclusiveness of the vast differences of people and their wants and needs. As one example, the experience and needs for safety are different for womxn, people who are LGBTQIA+, children, and mature adults than they are for middle-aged able-bodied men. Designs have not grown to reflect the safety needs and experiences of womxn, people who are LGBTQIA+, children, and mature adults as common practice, and it shows in how these groups of people use – or more importantly, do not use – public spaces.

Inclusive designs consider the wants and needs of all its users and do not prioritize a one-size-fits-all approach. To be inclusive in design helps the community and culture to embrace the needs of a great range of people, and therefore have spaces for a greater range of people to be seen, acknowledged, celebrated, and included based on their needs and experiences.

1.2 Understanding Equity, Inclusion, and Other Terms

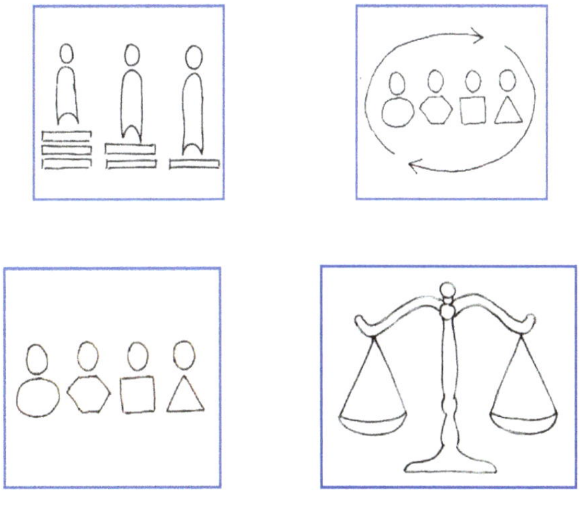

Understanding equity and inclusion and other terms as they apply to design helps in understanding the practices that can aid in improving their applications. The understanding and use of these terms specifically focus on design practices.

1.2.1 Equity

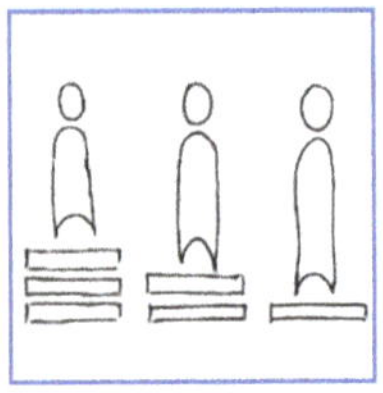

Equity and equality are two different terms with two different meanings. Image 1.7 illustrates the difference between the two terms visually.

To apply design equally fails as a false premise that everyone's needs in design have equal applications. However, as there is a great range of needs for people based on their health, abilities, sexuality, gender, age, race, language, income, and culture, the needs of designs to address improved spaces are about equity and giving people who need more or differently those needs in designed spaces.

Image 1.7 **The difference between equality and equity in visual terms. As people need different design approaches, equity addresses the giving people what they need to have a similar and balanced experience in a space.**

1.2.2 Inclusion

Inclusion addresses the intentional and conscious effort of making sure people of a variety of health, abilities, sexuality, genders, age, race, language, income, and cultures participate in the project. Intentional, conscious, and caring inclusion in design improves spaces for a greater community of people to celebrate, live, and grow together.

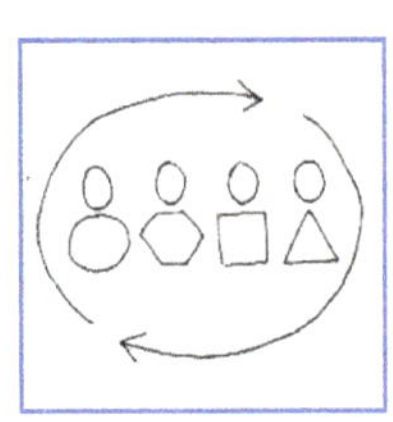

1.2.3 Diversity

Diversity of people in green design is based on their health, abilities, sexualities, genders, age, race, language, income, and cultures. Diversity of people is not typically celebrated in traditional public spaces in the United States, as the tradition of public urban and suburban spaces was aligned by and for people who are white, middle-aged, middle- and upper-class men (Cohen et al., 2021; Nagendra and Sen, 2022). The effects of the lack of diversity in the people who created the designs of spaces still affect present spaces today as some spaces are decades or hundreds of years old and do not consider the wants and needs of a greater range of people.

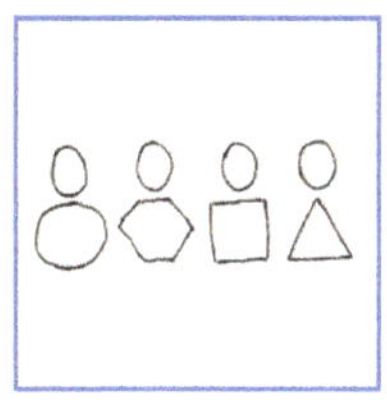

1.2.4 Justice

Justice is interesting, as it reminds me of redlining areas and maps and zoning practices of decades ago and its present-day effects. Further, I am reminded of the wrongful alteration of community boundaries to favor a political party. The effects of these on voting, policies, administration, and public space are felt not only from the past but also in present-day practices. These are only some of the issues of justice in design. Justice in designs shows in both large-scale and small-scale spaces.

One focus in this book is the lack of justice in a disproportionately low amount of open green community spaces in Black, Brown and underserved communities, as well as in areas of people with low income. The lack of healthy green spaces that were not designed into these communities affects both the physical health and emotional health of people in these communities. One example is poor air quality due to the lack of green spaces is physically damaging for people's breathing, and the lack of green community spaces for gathering with friends and neighbors is challenging people's emotional health (American Chemical Society, 2009; Raanaas et al., 2011; Wells and Evans, 2003; Wolf and Flora, 2010; World Health Organization, 2016).

1.3 Components of Equity and Inclusion – Understanding Health, Abilities, Sexuality, Gender, Age, Race, Language, Income, and Cultures

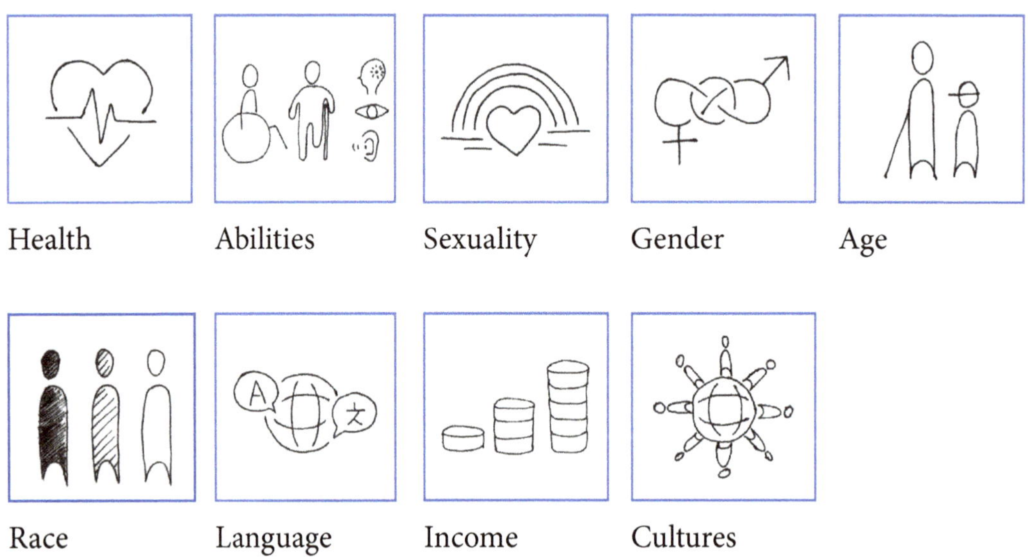

Understanding the use of the terms health, abilities, sexuality, genders, age, race, languages, income, and cultures in design helps explain the practices that can be used for improving green design for equity and inclusion.

1.3.1 Health

Both physical and emotional or psychological health are applicable to green design.

In this book, emotional health is a more leading focus, with some specific directions for physical health as it affects breathing with air quality and safety from flooding. Using green design can improve air, water, and soil qualities. Green design can also have psychological benefits for people, as stated in multiple studies (American Chemical Society, 2009; Raanaas et al., 2011; Wells and Evans, 2003; Wolf and Flora, 2010; World Health Organization, 2016).

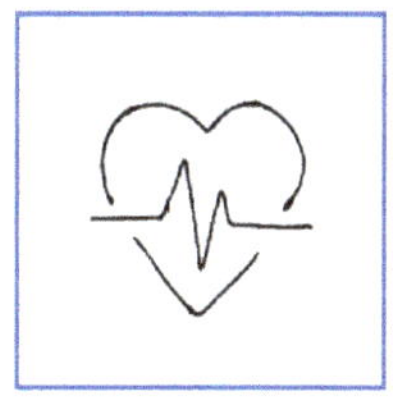

1.3.2 Abilities

This describes, more specifically, different abilities. The terms "disability" and "special needs" are replaced with the term "different abilities," a chosen phrase by people with different abilities including physical, intellectual, or cognitive. This term is inclusive of people with different physical abilities such as walking, seeing, and hearing, as well as different learning or cognitive abilities such as people on the spectrum or with autism.

There are physical design features that are beneficial for people who have different visual, hearing, or walking abilities. There are also physical design features that are beneficial for people with different cognitive abilities.

1.3.3 Sexuality

The present acronym used for sexuality is LGBTQIA+ which is an acronym for lesbian, gay, bisexual, transgender, queer or questioning their gender, intersex, asexual, and their allies. A more recent acronym in use is LGBTQIA2S+ which adds two-spirit to the previously described acronym.

1.3.4 Gender

More specifically in this context, designing public and private spaces for womxn and people who identify as a gender other than womxn or man.

1.3.5 Age

Age addresses people who are mature or advanced in age, as well as people who are young, such as children and teenagers. Green design has different uses and needs for people of different ages.

1.3.6 Race

Race includes people who are Black, Brown, Indigenous, people of color, mixed race, and in underserved communities. The term BIPOC will not be used in this book, as noted earlier in Chapter 1 about the issues of acronyms, who creates them, and not seeing the issues and needs of people.

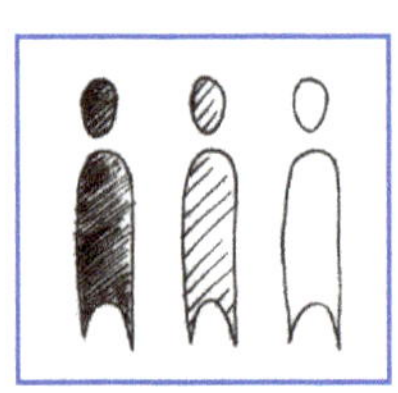

1.3.7 Language

Language includes people who are in locations where the official, first, or primary language of the land or territory is not the individual's first language. Additionally, there are areas where there is no official language, as well as where there are multiple language practices.

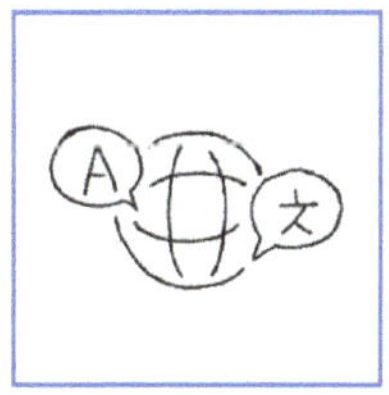

1.3.8 Income

Income specifically addresses people who have low incomes. Low income does not only include an individual or family who is at or below the poverty level; it includes people who are not paid the cost of living in an area.

1.3.9 Cultures

Culture is about looking at people of different values, philosophies, beliefs, ethnicities, practices, and traditions from around the globe who may not be living in their native area. Additionally, people of mixed cultures are also included in this term.

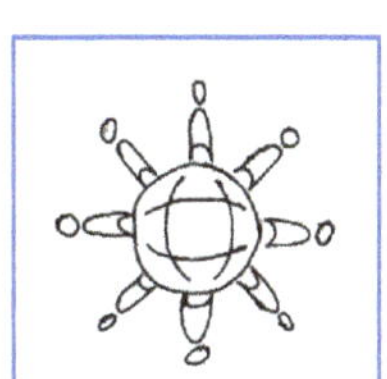

1.4 Equity and Inclusion in Green Design

Images 1.8–1.10 Redesigning spaces for people that are inclusive include designs for womxn, people with a wide range of abilities, and people who are mature.

Exploring equity and inclusion in green design – as well as the further focus of health, abilities, sexuality, genders, age, race, language, income, and culture – can be quite a large scope. Further, issues of the use of spaces, economic diversity, and multiple other issues arise while designing public and private spaces. This book uses design ideas and case studies to address green design solutions for improving spaces to address equity and inclusion for their health, abilities, sexuality, genders, age, race, language, income, and culture.

In broad strokes, this book explores inequity and a lack of inclusion as it shows in green design through the following.

1. Lack of agriculture and access to healthy food (food deserts) in low-income communities (Chapter 2).
2. Lack of safe spaces due to flooding and stormwater in communities (Chapter 3).
3. Lack of green spaces to improve air quality for health (Chapter 4).
4. Lack of green spaces for recreation and gathering in lower economic areas and communities which have people who are Black, Brown and from underserved communities, as well as in low-income communities (Chapter 5).
5. Lack of open green spaces that are safe and inclusive for diverse community members due to materials, lighting, design, use, programs, scale, and wayfinding (Chapter 6).

Image 1.11 Flooding disproportionately affects areas of people with low income.

1.5 Examples of Inequity and Lack of Inclusion Show in Existing Green Spaces

In looking at some details for examples of understanding design, inequity, and a lack of inclusion, they can be seen by a lack of seating for people who are shorter, such as children, and shared spaces for people who are taller and shorter. One detail is that bench heights are made for men of average height rather than for people who are pregnant, womxn of average height, people with different needs for comfort, and children. Further, the use of colors, shapes, forms, and plant materials of other cultures can be lacking in spaces, as well as the use of multiple languages in signs for inclusive communication. The height differences of signs for readability are for an average height man, not womxn, children, or people using wheelchairs or other mobility devices.

In the United States, the narratives and stories that are shown through spatial design have been and presently are those of Caucasian males, as well as middle to upper Caucasian culture (Cohen et al., 2021; Nagendra and Sen, 2022). In the United States, the fields that design spaces – including green design, landscape architecture, planning, architecture, engineering, and interior design – are not by majority populated by womxn or people who are Black, Brown, and from underserved communities (Zippia, n.d.).

The 2021 statistics for designers in the United States illustrate the following.

- 56.9% of designers are men, and 43.1% are womxn.
- 70.2% of designers are Caucasian, 10.6% are Hispanic or Latino, 10.4% are Asian, 5% are unknown, 3.4% are Black, and 0.4% are American Indian and Alaska Native.
- 13% of designers identify as LGBTQIA+.

Image 1.12 **Creating public seating for comfort for people who are mature includes designs with backs and side support.**

The lack of diversity in practitioners creates designs that not only follow traditionally able-body male-dominated designs, but also traditions of Caucasian practices that neglect to consider other interests and practices – such as those of other cultures, the needs of those with different abilities, different genders, sexualities, and low income. Expanding the design fields to be welcoming, encouraging, supportive, and open to diverse practitioners supports change in spatial design.

The U.S. legislation known as the Americans with Disabilities Act (ADA) was started for public accommodations in the 1980s and was signed into national law in 1990. Since then, updates to the law continue to be signed into law. Design inclusions of the ADA policy include requiring ramps and signage for those with different walking and visual abilities. In practice, I have worked at offices and with clients who follow the letter of the ADA law with little consideration for the effects on the people who need these design practices, let alone more than the law and design toward the spirit of updated practices for inclusivity. As a result, walkways and signage are designed as secondary or afterthoughts, which can be seen in some minimal design considerations in site design.

When a design prioritizes a one-size-fits-all approach, true inclusivity is lost. The ADA falls short in creating effective design, which goes beyond basic compliance to real work of thoughtful inclusive projects. True inclusivity requires designs that consider the wants and needs of all users who are on site, as well as looking at those who are not on the site and asking why they are not there. If the answer is because it is not an inclusive space, then design may be – in part – at fault; therefore, it must also be part of the solution.

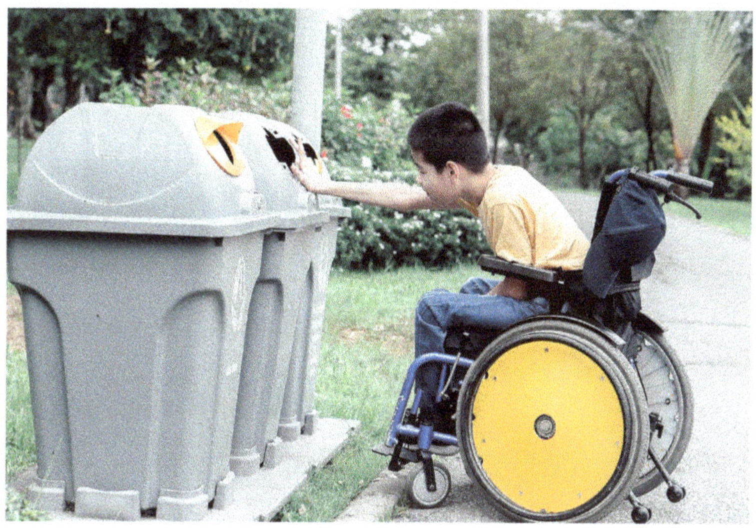

Image 1.13 **The public trash cans in this space are not accessible by people who use wheelchairs, which creates an inequitable space for this individual.**

Equity and inclusiveness can and should be the primary focus of design practices. In my university courses, I require students who can walk, once a semester, to spend five to ten hours in a wheelchair to experience spatial design considerations from the perspective of someone who cannot walk. This exercise is not for understanding a lifestyle or even a day for people who use a wheelchair as a lifestyle. This exercise for design students is to experience spaces, walkways, blocked paths, materials, views, textures, pace, lighting, lack of ability to experience shade under a tree in a grassy area, signage at uncomfortable heights, and much more in public and private spaces.

The exercise teaches students to design for inclusiveness for those with different mobility. Each class member then writes an essay and develops a design of a space that is beyond the letter of the law but includes the spirit of inclusion with an intentionally effective design of equity for those who use a wheelchair to be able to move independently and comfortably. The improved designs consider sidewalk zones wide enough for an individual in a wheelchair to navigate with a friend who can walk, while talking next to them, shade structures that are wheelchair accessible, trash cans with covers (as the scent coming from them is at the same height as the faces of wheelchair users), and much more.

1.6 Understanding Social Determinants of Health (SDOH) and Their Relationships to Health, Abilities, Sexuality, Gender, Age, Race, Language, Income, and Cultures, and Addressing Ideas and Design Solutions

Image 1.14 **The social determinants of health (SDOH) and their relationship to health, abilities, sexuality, gender, age, race, language, income, and cultures.**

The social determinants of health (SDOH) were developed by the U.S. Centers for Disease Control and Prevention (CDC) to address non-medical factors that influence health. There are a range of conditions that influence people's health in their daily lives, including economic policies and systems, growth agendas, social norms, social policies, and political systems.

Along with health equity and health literacy, the SDOH are used to create national objectives with the data research compiled by the CDC. The five components of the SDOH include healthcare access and quality, education access and quality, social and community context, economic stability, and neighborhood and built spaces. The applications of these include safe housing, transportation, and neighborhoods; addressing polluted air and water; and access to healthy foods and physical opportunities.

Addressing differences in SDOH accelerates progress toward health equity, a state in which every person has the opportunity to attain their highest level of health. SDOH have been shown to have a greater influence on health than either genetic factors or access to healthcare services. For example, poverty is highly correlated with poorer health outcomes and higher risk of premature death.

SDOH, including the effects of centuries of racism, are key drivers of health inequities within communities of color. The impact is pervasive and deeply embedded in our society, creating inequities in access to a range of social and economic benefits – such as housing, education, wealth, and employment. These inequities put people at higher risk of poor health.

CDC is coordinating efforts to focus its resources on the areas where federal public health investments can make the most difference. For example, CDC's Racial and Ethnic Approaches to Community Health (REACH) focuses on reducing high rates of chronic diseases for specific racial and ethnic groups in urban, rural, and tribal communities. Since 1999, the program has worked across sectors in racial and ethnic minority communities to reduce tobacco use, improve access to healthy foods, change the built environment to promote physical activity, and connect people to clinical care.

CDC (n.d.)

Image 1.15 **Public green spaces in communities are important determinants of health for people who are marginalized and underrepresented.**

The connection of using green design – such as those illustrated throughout this book – to address health, abilities, sexuality, gender, age, race, language, income, and cultural equity is an important role. Communities rarely have occasions to create completely new neighborhoods and spaces. As populations increase, retrofitting locales is a common practice to address growing societies. Therefore, being creative in altering people's approach to necessary green design for health can support equity and inclusion in all spaces where people live, work, travel, or spend any time.

There are some designs in this book in which the SDOH approach of creating healthy built surroundings to address health equity is applied. For example, though there may not be a specific design feature used to address racial inequity, health equity is created by designing green spaces in Black, Brown, and underserved communities. This applies the SDOH that address the practice that creating healthy green spaces affects one's health more than people's race, culture or genetics.

In extending the applications of the SDOH in which design of spaces is an important determinant of health, this book also applies healthy greening and other components for health equity. This is done to invest in creating spaces in which health is prioritized through spatial design for people who have been or are marginalized through a range of areas such as health, abilities, sexuality, gender, age, race, language, income, and culture. The creation of healthy spaces for people who are and have been marginalized is an extension and application of the SDOH.

1.7 Issues in Green Design Related to Health, Abilities, Sexuality, Gender, Age, Race, Language, Income, and Cultures, and Addressing Ideas and Design Solutions

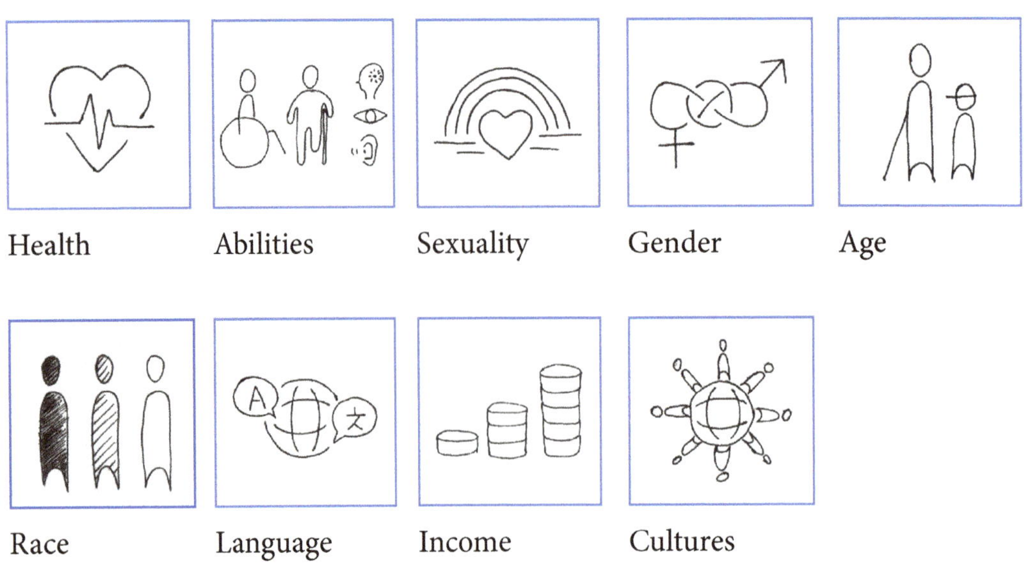

There are a range of issues that show a lack of inclusion and equity in green design and public spaces. Inequities and lack of inclusion are shown in small subtle approaches, as well as large overarching practices and conditions. This chapter is an overview to understand terms and how spatial design is not equitable or inclusive. The upcoming chapters illustrate designs that shift thinking on equity and inclusion, as well as providing articles and case studies that illustrate challenges and solutions toward equity and inclusion.

In Chapter 7, there are lists that illustrate green design issues and solutions of health, abilities, sexuality, gender, age, race, language, income, and culture in spaces. The lists in Chapter 7 are applicable not just to the sites shown in this book but also to sites where people walk and live every day. People may or may not realize the indicators of the lack of inclusion and equity in sites – and this is one of the points of the book. Realizing and understanding the preferential issues are necessary steps to changing designs, design approaches, education, policies, and practices that continue to contribute to inequity and exclusive designs and spaces.

It is not only educators, designers, policy makers, and administrators who can benefit from understanding equitable and inclusive green design approaches – it is members of the community. It is through their voices and actions that creating change is possible. The more people from diverse backgrounds, knowledge, and skills become aware of how design can be equitable and inclusive, the more it can change and be used and expanded.

1.8 Moving Forward for Equitable and Inclusive Design

There are countless approaches beyond green design to address issues of equity and inclusion that design cannot. Creating local, regional, state, and national policies and training – as well as putting people who are womxn; people who are Black, Brown, and from

Image 1.16 Having diversity of people with different needs and interests in administrative roles can help with understanding, advocating, and changing policy and culture.

underserved communities; and people who have different abilities and who are from different cultures and economic levels in administrations and roles where they can effectively cause change – is vital. Further, removing people who do not practice, complicate, and hinder equity and inclusion from roles where they continue to practice discrimination subtly and openly is required in order for change to happen.

There are people in a great range of fields such as academia, public policy, administrations, governing agencies, governing boards, community agencies, the arts, developers, educators, and much more who have important roles in policies and design of public spaces. Causing change in details as well as in big broad strokes with inclusion and consideration requires addressing the layers of people and policies involved.

There is not a single space where inclusion and equity of people based on their health, abilities, sexuality, gender, age, race, language, income, or culture do not belong.

In Chapters 2–6, there are multiple examples of applying equitable and inclusive design in a variety of sites, such as sites that provide healthy food, sites that flood, sites that address a healthy society including air quality, sites that are community spaces, and sites for public comfort and safety. Applying equitable and inclusive design, both for new sites and retrofitting, is required for moving forward in site design, design practices, planning processes, policy making, community planning, public workshops, and education of all of these.

Equality and Inclusion in Green Design: Inclusive Approaches for Designing Healthier Work and Living Spaces focuses on seeing different spaces for the greening they can be, as well as teaching people to understand how to see unique ideas toward equity and inclusion in design. Green design is diverse and ever-changing and can be used for addressing some of the inequity and exclusion that is present in private and public spaces.

Images 1.17–1.21 Designs throughout this book explore spaces redesigned to address equity and inclusion of people in exterior and interior spaces.

As there are more modalities and the needs and locations where people now work, making sure that multiple types of spaces are designed for the diversity of people's success is more relevant than ever. This includes designs for spaces that are urban, suburban, rural, towns, offices, commercial spaces, warehouses, plazas, homes, studios, and more.

This book gives readers a way to not only understand greening, but also to understand how to see and apply greening to their site – specifically toward creating equitable and inclusive design for the strength of their business, health center, university, education center, home, street, neighborhood, community, and region. Creating both new spaces and retrofitting spaces to specifically address issues of a long unjust past is one goal of this book. Realizing that it is not solely an unjust past which is being practiced in present day is another goal. Finding solutions that do not settle for merely improving spaces – but rather remove design inequities and exclusion – is needed for acknowledging the past, it is also for being aware of the present, and changing the present and future directly.

References

American Chemical Society. (2009, July 23). Stop and Smell the Flowers – the Scent Really Can Soothe Stress. *ScienceDaily*. Retrieved 26 April 2023, from www.sciencedaily.com/releases/2009/07/090722110901.htm

CDC. (n.d.). Retrieved 2 January 2025, from https://www.cdc.gov/about/priorities/why-is-addressing-sdoh-important.html#cdc_generic_section_3-public-health-actions-that-affect-sdoh

Cohen, D. A., Williamson, S., & Han, B. (2021). Gender Differences in Physical Activity Associated with Urban Neighborhood Parks: Findings from the National Study of Neighborhood Parks. *Women's Health Issues: Official Publication of the Jacobs Institute of Women's Health*, 31(3), 236–244. https://doi.org/10.1016/j.whi.2020.11.007

Kratz, J. (2024). The Little Known History of DEI and Why it's Critical to its Survival. *Forbes*. Retrieved 20 February 2025, from https://www.forbes.com/sites/juliekratz/2024/12/29/history-of-dei-why-it-matters-for-the-future/

Nagendra, H., & Sen, A. (2022). Our Cities are Designed by Men, for Men. *The Third Eye and Science the Wire*. Retrieved 20 February 2025, from https://science.thewire.in/society/gender/our-cities-are-designed-for-men-by-men/

Raanaas, R. K., Horgen Evensen, K., Rich, D., Siostrom, G., & Patil, G. (2011). Benefits of Indoor Plants on Capacity in an Office Setting. *Journal of Environmental Psychology*, 31(1), 99–105.

Wells, N., & Evans, G. (2003). A Room with a View Helps Rural Children Deal with Life's Stresses. *Environment and Behavior*, 35(3), 311–330.

Wolf, K. L., & Flora, K. (2010). Mental Health and Function – A Literature Review. In: *Green Cities: Good Health*. College of the Environment, University of Washington. www.greenhealth.washington.edu

World Health Organization. (2016). *Urban Green Spaces and Health*. Geneva: WHO Regional Office of Europe.

Zippia. (n.d.). *Designer Demographics and Statistics in the US*. Retrieved 25 February 2025, from https://www.zippia.com/designer-jobs/demographics/

Chapter 2

Access to Healthy Food

Images 2.1–2.4 **Creating public interior and exterior sites as spaces for growing community food creates food security, as well as healthy food sources for a community of people.**

DOI: 10.4324/9781003475293-3

2.1 Introduction

In Chapter 2, we learn that exploring food access in designs – as well as seen through case studies and research about food access and clean drinking water – is important for food security. The practice of creating healthy green design solutions at both site scale and community scale, interior and exterior, hydroponic and soil based, allows unique site-specific design interventions for a myriad of sites.

Unique solutions for food security can be designed with food-based plants in public spaces, which allow people who want or need healthy food to have a greater access to it. Street trees, extended sidewalk greening, street medians, green walls, shipping container gardens, alleys, warehouses, and other unique solutions are available options to use for food-based designs in community spaces.

Creating spaces with the intentionality of food and water security for marginalized populations encourages engaging the community toward healthy solutions. Along with public spaces for food growth, the opportunity for people to work in the green field creates strong community growth toward a healthy sustainable future. Training people in the community to learn green maintenance, an employable skill, can be a source of pride for the district. The food produced in these sites can be given to the people in the community, shared at public events, or donated to local kitchens.

Pride, health, equity, and inclusion become sources of short-term and long-term success when food security is acknowledged, seen, understood, and addressed for the public in community spaces. Food security is a basic need in order to be successful in the growth of individuals, as well as a community, region and culture. Using public resources for healthy food illustrates equity for people who have been and continue to be underserved.

2.2 Urban Community Green Design

At the community scale, the drawing in Image 2.6 illustrates layers of green design approaches for access to growing food and equitable and inclusive spaces for marginalized populations in both public and private environs. Creating a pedestrian scale space means using design approaches that are specifically aimed at comfortable and safe areas equitably and inclusively for people to sit, walk, run errands, use public transportation, and share space easily.

This design in Image 2.6 includes wide sidewalks with zones for business fronts, signage, seating, green design, walking, scooter and bike parking, expanded tree wells for gardens, trash cans, public comforts, and crosswalks. Pedestrian-scale warm lighting is used to create a safe space for womxn, members of the LBGTQIA+ community, and children. Multiple styles of seating options for people with different health abilities, as well as people of different heights, are placed at public transit locations as well as along the sidewalks. Accessible public transit allows for users with a range of incomes and

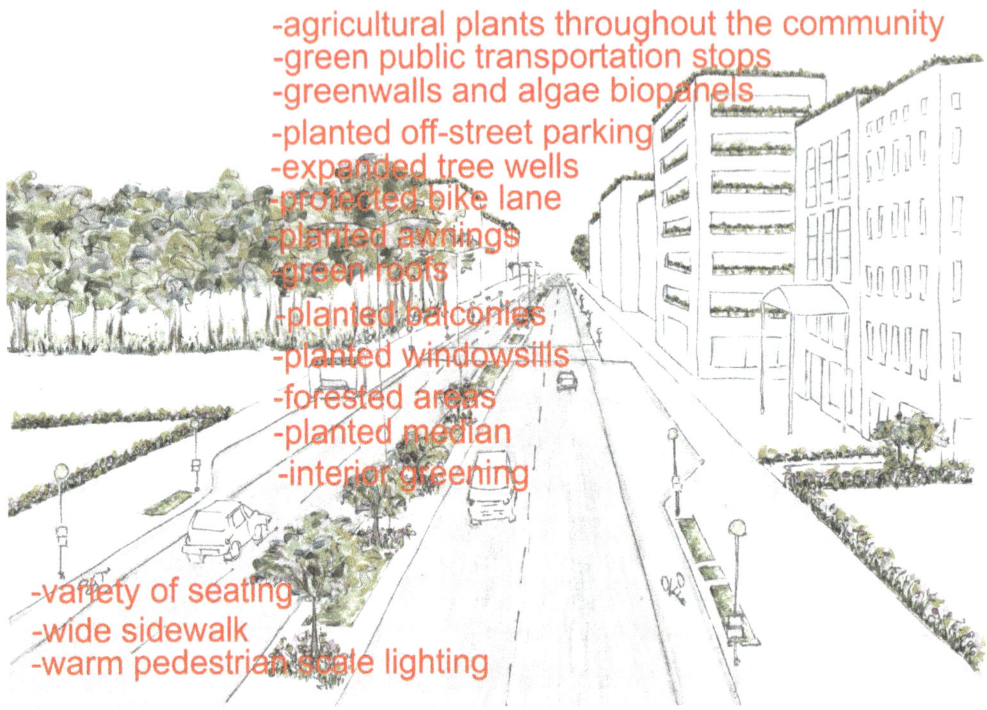

Image 2.5 Using public space for growing food in urban areas creates opportunities for people within the community to share food for all the people of the community who want or need it.

abilities – as well as large households with only one vehicle – to have consistent access to navigate throughout their community with ease and safety.

A protected bike lane on the street encourages shared spaces for multiple users. The crosswalk is designed with colorful art created with ecological paints that reduce air contaminants, created by local and regional artists to celebrate culture and change. This improves the shared spaces for air quality health for the community.

Planted designs are used in the parking lots, which are off the street and have living green planted fences for improving air quality. Permeable pavers in the parking lots allow stormwater to filter through the ground, reducing the amount of water that goes to the overtaxed city sewer plants.

Greening with planted designs are used in small and large scales in this urban community with expanded street tree wells for gardens, green balconies, planted rooftops, and algae biopanels on buildings. The green median in the center of the street is planted, and there are lights for safety and seating for comfort. The planted design throughout these layers not only improves air quality but also uses food-producing plants. Put in public spaces, these plants allow people who want and need food to have access to it, as well as creating job opportunities for community members to care for and distribute the food to community members, food pantries, education centers, and other community food-sharing programs for food security.

Image 2.6 Urban panoramic green design.

Using public design to create spaces where diverse users are welcomed requires healthy inclusive practices. Green design is developed on walking paths, at eye level, and above eye level in unique practices in this urban site with plants that produce food.

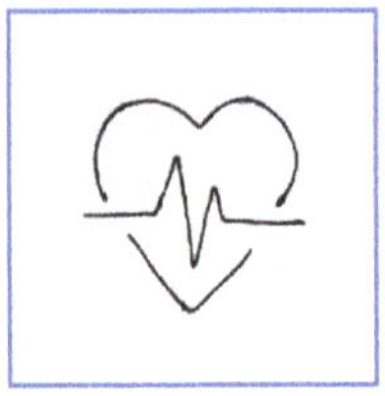

Health – The addition of green roofs, balcony plantings, an urban park, expanded tree wells, planters, algae biopanels on buildings, and planted bus stop covers improves air quality for easier breathing, as well as for emotional health. The plant selection produce food, which is shared throughout the community. Street murals use air-purifying paint, which absorbs and neutralizes chemicals, pollutants, and volatile organic compounds (VOCs) for improved air quality. The center median has a lush green design to improve air quality and create a space for stopping, sitting, or community gathering. The center area may be raised or at street height for safety of the pedestrians – as well as the drivers – to reduce glare from headlights of vehicles on the opposite side at night.

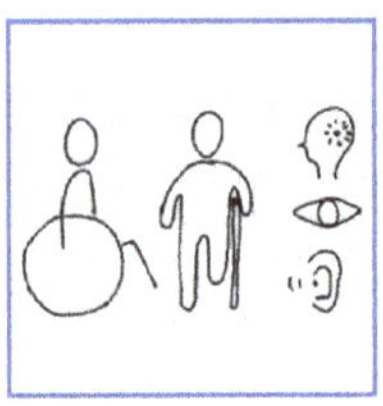

Abilities – The center median provides safe resting spaces for people who need a longer time for crossing. The street murals use colors visible to people with different visual abilities. Curb cuts, detectable rumble strips and warning pavers at corners, crosswalk melodies for walk and do not walk, and high-contrast signage improve navigation for people with different visual and walking abilities. A variety of seating options and spaces for people with different mobility skills are available. There are zones for pedestrians to move at their pace.

Sexuality – Consistent luminescent lighting at pedestrian scale along each block, as well as warm lighting, improves safety for the LGBTQIA+ community. The signage of local businesses welcomes and celebrates people of this community.

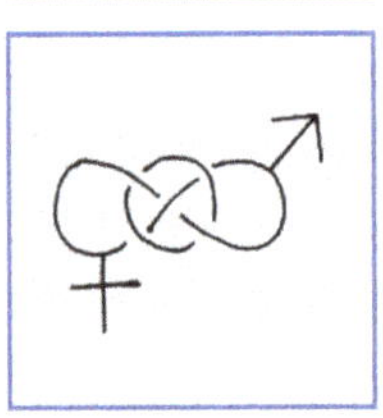

Gender – The center median allows resting or stopping spaces for caregivers with children and mature adults. Consistent luminescent lighting at pedestrian scale along each block, as well as warm lighting, improves safety for womxn in the community. Wider sidewalks are spaces for strollers and holding children's hands. Public transportation is routed for womxn and people who run errands and take routes with multiple stops. Designs, signs, and street names celebrate womxn.

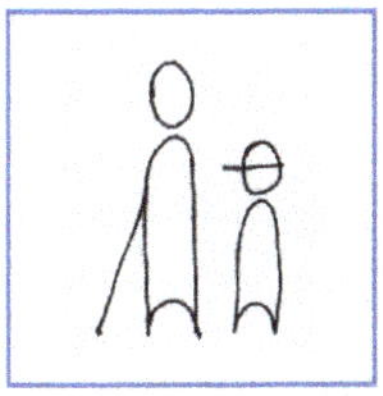

Age – The center median allows resting or stopping spaces for people who need time to cross the street. Curb cuts and safe crosswalks are at every street corner for people to safely cross streets at multiple locations in their route. This is important for people who walk throughout their day while running errands with children and people who are mature.

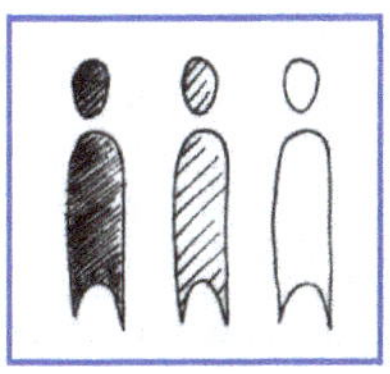

Race – Designs, signs, and street names celebrate people of different local, regional, national, and international communities.

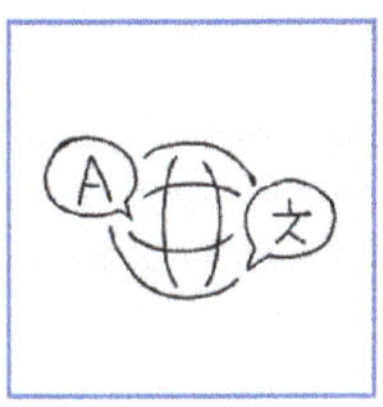

Language – Signs that use multiple local, regional, and international languages, as well as icons and images without words for those learning language, are used throughout the street at pedestrian heights.

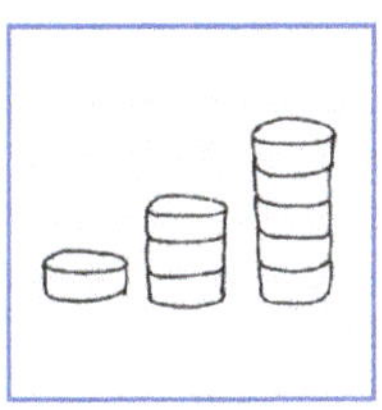

Income – Communities of people who have low income and have been disconnected by roadways are connected to their surrounding communities with a safe pedestrian greenbelt. The green median has food-producing plants, which are available to eat to support people with low income in the community who want or need healthy food. The care for the median can also provide jobs to people who want temporary or long-term jobs.

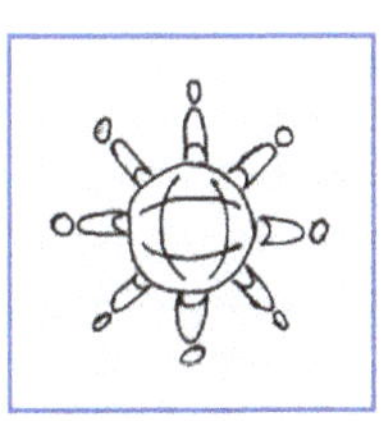

Culture – Food, designs, signs, and street names celebrate people of different local, regional, national, and international communities.

Creating a pedestrian-scale green design that creates equitable and inclusive spaces in an urban area involves layers of green design solutions.

Image 2.7 **Rooftop garden at the University of Warsaw Library, Warsaw, Poland.**

2.3 Rooftop Garden at the University of Warsaw Library, Warsaw, Poland

The rooftop garden at the University of Warsaw is one hectare in size. One hectare is the equivalent of 10,000 square meters, or approximately 2.471 acres. Opened in 2002, it is one of Europe's largest rooftop gardens. A cascading stream joins the upper and lower parts of the design.

Image 2.8 **The green area of the upper rooftop garden at the University of Warsaw Library, Warsaw, Poland.**

The lower area includes a fishpond where ducks have taken residence, as well as a series of granite sculptures by Ryszard Stryjecki on cosmological themes. The upper garden comprises four parts: the golden part (to the north), the silver part (to the east), the crimson part (to the south), and the green part (to the west). The colorful vegetation of each of the sections is reflected in the names of the areas. Individual sections of the garden are connected by paths, bridges, and pergolas.

2.4 Thammasat University Green Roof, Thailand

Thammasat University in Thailand also has a rooftop garden, which includes a solar roof, urban farming, and green public space. The 22,000 square meters (236,806 square feet) produce food for the campus community and beyond. Pollinator plants, as well as evergreens and other communities of plants and trees, round out this space to create an ecosystem that is irrigated with stormwater.

Universities are ripe with opportunities to develop spaces for teaching agricultural designs, as well as other green technology. Through example, multiple programs of study such as landscape architecture, horticulture, architecture, engineering, art, public art, sculpture, farming, agriculture, and other fields can work together for cooperative learning and expansion.

Image 2.9 **Rooftop garden at the University of Warsaw Library, Warsaw, Poland**

Image 2.10 **The golden area of the upper rooftop garden at the University of Warsaw Library, Warsaw, Poland**

Teaching inclusion at universities through example is not just words, but actions to help communities and people within them to find solutions to make healthy food security a right, rather than a luxury.

2.5 Urban Agriculture Green Design

Urban areas have unique sites that can be used for greening, such as underutilized spaces adjacent to or under roadways. The design in Image 2.12 uses an area under a roadway for urban agriculture. This is to create community food safety for people in the community who need or want healthy food.

Shipping containers are utilized for urban agriculture, with food being grown in and on the containers. The design stacks containers two high and uses a shipping container for storage of tools.

Rain barrels capture water from the site to use for irrigation, in addition to the stormwater and greywater – which comes from sources with mild impurities and includes water used in sinks, showers, and washing machines – of nearby sites. Permeable pavers are used throughout the site, which allows stormwater to filter through the ground, reducing stormwater runoff and reducing water that goes to the overtaxed city sewer plants.

Multiple styles of seating options for people with different health abilities, as well as people of different heights, are used throughout the space. Seating is for the community to share time, care for the space, and work with the site, encouraging a shared community space for people. Warm pedestrian-scale lighting through the space creates safety for people throughout the day and night, and allows night-time celebrations for the community.

Artwork created by the community of artists covers the area and can change throughout the year. Changing art encourages people to see the space with fresh eyes as art changes. Artwork on horizontal and vertical surfaces throughout the space, as well as on the ground, is a vast opportunity to celebrate the community through art. Using air-purifying paint, the paint helps to reduce VOCs in the air.

The urban community agricultural site is accessible for the people of the community to have access to growing food for the people who want and need it, particularly the underserved and marginalized populations. Accessible public transit allows users with a range of incomes and abilities, as well as large households with only one vehicle, to have consistent access to the community food source.

The planted design throughout these layers not only improves air quality but also uses food-producing plants. Put in public spaces, this design allows people who want and need food to have access to it, as well as creating job opportunities for community members to care for and distribute the food to community members, food pantries, education centers, and other community food-sharing programs for food security.

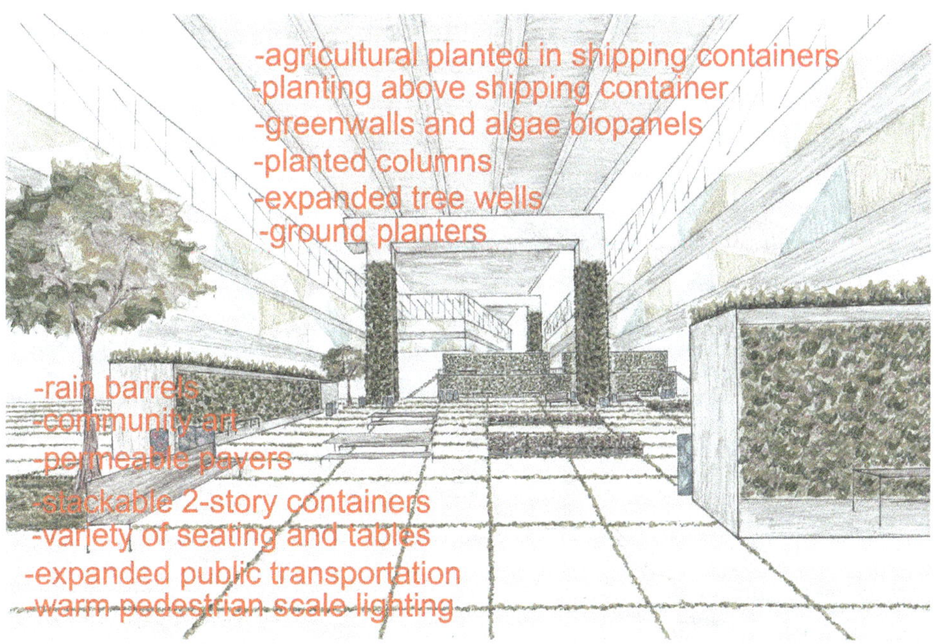

Image 2.11 **Under-utilized spaces, such as those under or nearby roadways, freeways, and highways, are utilized for community gardens, with shipping containers used for growing food.**

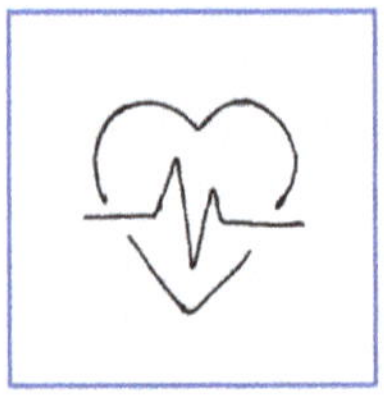

Health – The addition of shipping container gardens and greening under highways for urban agriculture gives a local source for healthy food. This design also improves air quality for easier breathing, as well as for emotional health. The plants selected produce food, which is shared throughout the community. Murals use air-purifying paint, which absorbs and neutralizes chemicals, pollutants, and VOCs for improved air quality.

Abilities – Planters of a variety of heights and shapes encourage users in wheelchairs to enjoy and care for the urban agriculture. The murals are of colors visible to people with different visual abilities. Detectable rumble strips and warning pavers are at the edge of the site, as well as crosswalk melodies for safe walking. High-contrast signage improves navigation for people with different visual abilities. A variety of seating options and spaces for people with different mobility skills are available.

Sexuality – Consistent luminescent lighting at pedestrian scale along each block, as well as warm lighting, improves safety for the LGBTQIA+ community. Signage of local businesses welcomes and celebrates the people of this community.

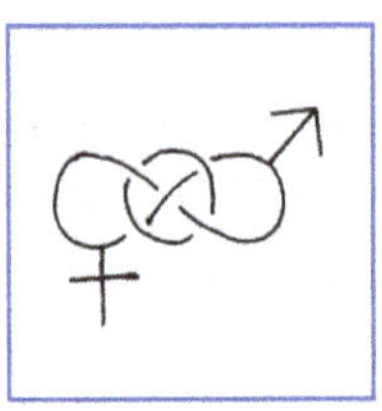

Gender – Consistent luminescent lighting at pedestrian scale within the space and walking to the space, as well as warm lighting, improves safety for womxn in the community. Wider paths are spaces for strollers and holding children's hands. Public transportation is routed for womxn and people who run errands and take routes with multiple stops with packages of the locally sourced food.

Age – Safe walking surfaces throughout the space are for people to safely enjoy the community space. This is important for people who walk throughout their day while running errands with children and people who are mature.

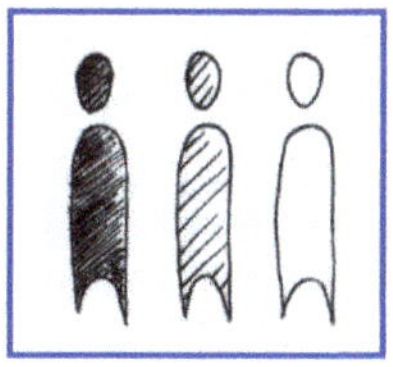

Race – Communities of people who are Black, Brown, and underserved who have been disconnected by roadways are connected to their surrounding communities with a safe pedestrian greenbelt. which includes urban agriculture. Murals are created to celebrate diversity in race. Designs, signs, and street names celebrate people who are Black, Brown, and from underserved communities.

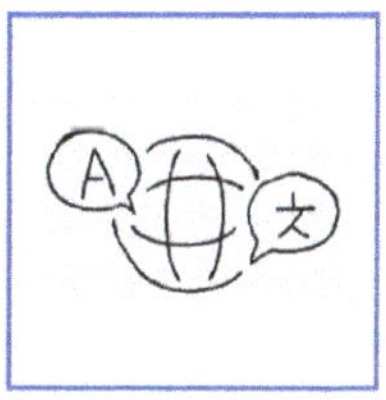

Language – Signs that use multiple local, regional, and international languages, as well as icons and images without words for those learning language, are used throughout the street at pedestrian heights.

Income – Communities of people who have low income and have been disconnected by roadways are connected to their surrounding communities with a safe pedestrian greenbelt. The green median has food-producing plants that are available to eat to support people with low income in the community who want or need healthy food. Care for the median can also provide jobs to people who want temporary or long-term jobs.

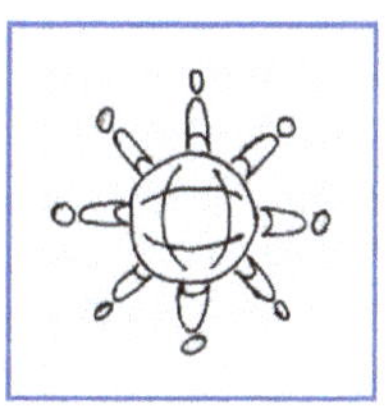

Culture – Food, designs, signs, and street names celebrate people of different local, regional, national, and international communities.

Highways and freeways have been intentionally used to divide communities, separating people of different races, income levels, and cultures. Creating an agricultural greenbelt reconnects communities with a space where they can celebrate diversity and share food.

Image 2.12 Urban agriculture under highways and freeways.

Urban areas can retrofit the underutilized areas under highways and freeways as local resources for urban agriculture. Shipping containers on one level, as well as stacked to two levels, are used.

Images 2.13–2.14 **Food insecurity, including lack of access to healthy water, disproportionately affects womxn and marginalized groups of people.**

2.6 Improving Water Security Helps Reduce the Gender Gap in Mexico City

Womxn and people in marginalized groups are disproportionately affected by water scarcity and the lack of water for consumption (Bosch et al., 2021). Womxn often play a key role in collecting, storing, distributing, and overseeing water. The majority of water carriers in low-income and middle-income countries are womxn and girls. In Mexico City, in homes that do not have water tanks, womxn are usually responsible for obtaining, storing, and addressing the use and reuse of water. One report estimates that in some areas, womxn spend up to an entire working week on water-related activities for the household.

The time womxn invest in unpaid work is multiplied when water is scarce and they are primarily routinely responsible for getting a

Images 2.15–2.16 **In Mexico, womxn are often responsible for obtaining, storing, and addressing the use of water for the entire household. This unpaid work contributes to inequitable income for womxn.**

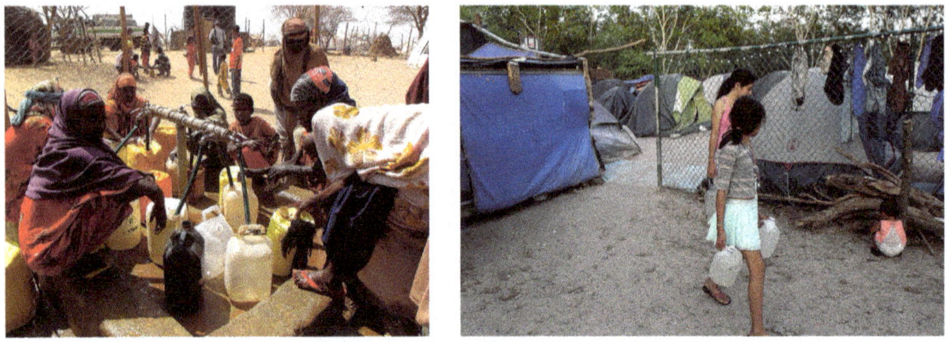

Images 2.17–2.18 **Womxn in these areas often spend the majority of their time in unpaid work, creating 17% of the workforce as womxn – continuing the inequity in work, education, and income.**

household supply of water. The disproportionate role womxn have in housework and childcare with the additional water sourcing results in an inequitable time for individual, educational, and professional growth and training for womxn and girls. Not unsurprisingly, womxn are underrepresented in water, sanitation, and hygiene sectors – creating only about 17% of the workforce.

One program in Mexico City is looking at rainwater harvesting systems to help address the social vulnerability of womxn. In Mexico City, 85% of the water comes from compromised sources. The water scarcity in Mexico City mostly affects communities with the highest number of marginalized people.

To help mitigate the water scarcity and inequity of womxn, the Mexico City administration in 2019 put into practice the Rain Harvest program for homes. The program builds rain harvesting systems that capture rain from

Images 2.19–2.20 **In Mexico, womxn are learning techniques for water harvesting to capture rainwater for homes and drinking (when further treated).**

the roofs, which is then stored in cisterns. The water is used for household purposes and – if further treated – for drinking.

The program prioritizes households headed by womxn, single mothers, Indigenous people, older adults, and people with disabilities in an effort to improve equity. As of 2021, more than 13,000 households in which womxn are the heads of the household have benefited from this program. Further, the training for the operation and maintenance of the system promotes equal distribution of tasks between womxn and men.

The results of these rain-harvesting systems are that 30% less time is spent by womxn on household water supply, with five to eight months of water coverage provided with the use of this system.

Nature-based solutions can help make water more available, more inclusive, and more equitable. Minority groups, including womxn, are often left out of the planning and process of water security. Green infrastructure goals are to protect, restore, and oversee natural ecosystems to address societal challenges, including water scarcity.

Green infrastructure projects such as these seek to develop solutions for the notable inequity of gender. It requires that cities employing these green solutions intentionally develop inclusive participation in the planning process, as well as when applying the solutions.

2.7 Agriculture in Public and Shared Spaces, Green Design

An alley of mixed private and public space is an opportunity to grow urban agriculture for the community that lives in and near the area. Image 2.22

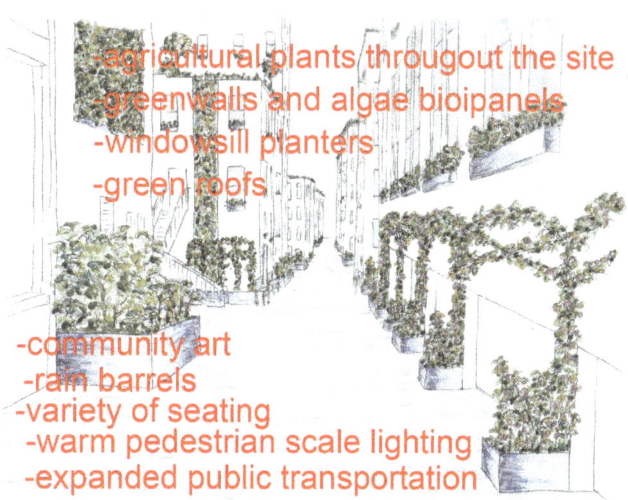

Image 2.21 **Alleys throughout communities are retrofitted as safe spaces for the community members with access to food for people who want or need it.**

uses healthy food and other plants that are grown vertically and horizontally throughout the space, including planted green roofs, green walls, algae biopanels, windowsills, and planters throughout the alley. Plants that produce food can include herbs, potatoes, lettuce, tomatoes, beans, peppers, and a variety of other edible options. Non-edible plants such as trees, shrubs, and plants create year-round green design in the community space to improve air quality, as well as the emotional and physical health of the community residents.

Rain barrels capture water from the site to use for irrigation, in addition to the stormwater and greywater of adjacent buildings. Permeable pavers are used in areas that allow stormwater to filter through the ground, reducing stormwater runoff and reducing water that goes to the overtaxed city sewer plants. Artwork created by the community of artists covers the area and can change throughout the year, celebrating local and regional cultures. Using air-purifying paint, the paint helps to reduce VOCs in the air.

Multiple styles of seating options for people of the community with different health abilities, as well as people of different heights, are used throughout the space. Warm pedestrian-scale lighting through the space creates safety for people throughout the day and night, and allows night-time celebrations for the community. Providing nearby accessible public transit for people to bring food home allows for users with a range of incomes and abilities.

The community-planted design throughout the site not only improves air quality and the emotional health of the community; it also provides food for people who want and need it. Put in public spaces, this design creates job opportunities for community members to care for and distribute the food to community members, food pantries, education centers, and other community food-sharing programs for food security.

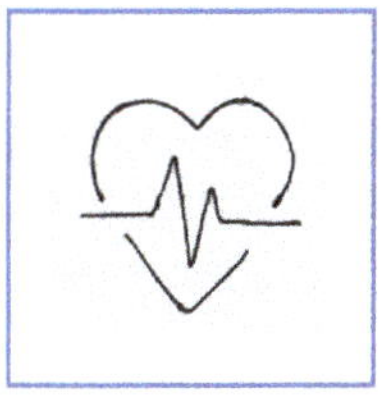

Health – The addition of planters, on windowsill for greening, vertical greening, and rooftop gardens creates a local source for urban agriculture. This design also improves air quality for easier breathing, as well as for emotional health. The plants selected produce food, which is shared throughout the community. Murals use air-purifying paint, which absorbs and neutralizes chemicals, pollutants, and VOCs for improved air quality.

Abilities – Planters of variety of heights encourage users in wheelchairs to enjoy and care for the urban agriculture. The murals are of colors visible to people with different visual abilities. Detectable rumble strips and warning pavers are at the edge of the site, as well as crosswalk melodies for walking. High-contrast signage improves navigation for people with different visual and walking abilities. A variety of seating options and spaces for people with different mobility skills are available.

Sexuality – Consistent luminescent lighting at pedestrian scale along each block, as well as warm lighting, improves safety for the LGBTQIA+ community. Signage of local businesses welcomes and celebrates people of this community.

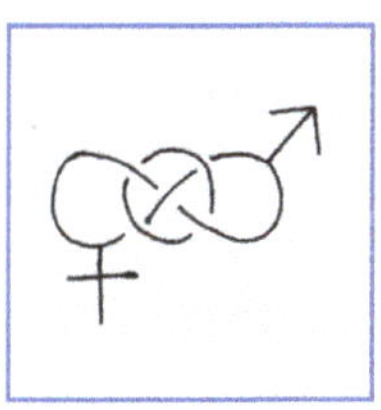

Gender – Creating a source of healthy food that is close by for womxn who work at home is an inclusive choice for their time and health. Consistent luminescent lighting at pedestrian scale within the space and while walk to the space, as well as warm lighting, improves safety for womxn in the community.

Age – Safe walking surfaces throughout the space are for people to safely enjoy the community space. This is important for people who walk throughout their day while running errands with children and people who are mature.

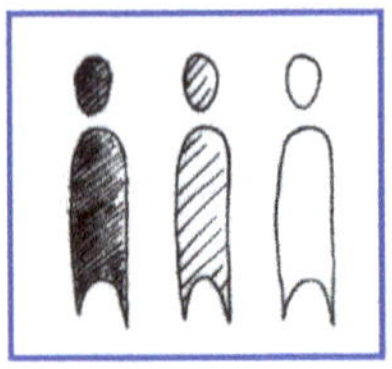

Race – The green alley addresses food deserts (areas that do not have local healthy food sources), which have been created predominantly in communities with people who have low incomes as well as people who are Black, Brown, and from underserved communities. Murals are created to celebrate diversity in race. Designs, signs, and street names celebrate people who are Black, Brown, and from underserved communities.

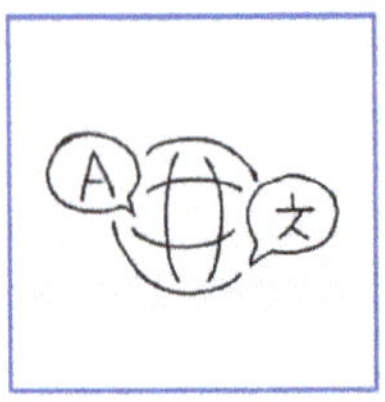

Language – Signs that use multiple local, regional, and international languages, as well as icons and images without words for those learning language, are used throughout the street at pedestrian heights.

Income – The green alley addresses food deserts (areas that do not have local healthy food sources), which have been created predominantly in communities with people who have low incomes as well as people who are Black, Brown, and from underserved communities. The local food is available to eat to support people with low incomes in the community who want or need healthy food. Care for the median can also provide jobs to people who want temporary or long-term jobs.

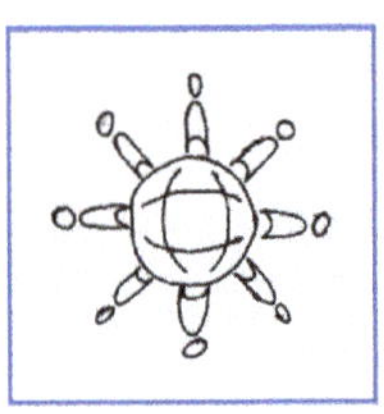

Culture – Food, designs, signs, and street names celebrate people of different local, regional, national, and international communities.

Alleys are underutilized areas where green design for the community, as well as services needed for the buildings, can be shared. An agricultural greenbelt can be a space to celebrate diversity and have changing agriculture.

Image 2.22 **Agricultural design in urban alley.**

This green alley design serves both as agricultural land and for connecting a community through an underutilized area. Serving for both greening and services, the alley is activated with greening on the walkway and windowsills, at eye level and rooftops. This design allows people who are outside to have healthy food, as well as see green design from their windows.

Image 2.23 **Farm for the City in Philadelphia, Pennsylvania, was installed in 2018 across the street from City Hall.**

2.8 Farm for the City, Philadelphia, Pennsylvania

An art-as-farm pop-up garden was located in Thomas Paine Plaza, across the street from the City Hall in downtown Philadelphia, Pennsylvania, in the summer of 2018 (ASLA, 2021). Farm for the City was a 4,000 square foot pop-up farm in downtown Philadelphia created by the Pennsylvania Horticultural Society.

Image 2.24 **The pop-up farm produced food for the community throughout the growing months of 2018.**

Image 2.25 **Throughout the time of Farm for the City, chefs prepared food for more than 8,600 unhoused, at-risk, and low-income people. Educational workshops for the community provided information on growing healthy food, as well as techniques for preparing meals with the freshly grown produce.**

The community garden addressed food insecurity for the unhoused, at-risk, and low-income population, with goals of helping to make healthy food more accessible, educating people about the food desert area, and highlighting how community gardens strengthen neighborhoods.

The space took one week to build with the selected materials chosen for their ability to be re-used after the completion of the pop-up, such as a custom light-weight soil mix for sustaining plant growth without heavy soil weight. Recycled chairs, tables, and umbrellas were added to the plaza. At the

Image 2.26 **Community members were able to work together to create resources for addressing food security in the short term with the installation, as well as in the long term with policies and practices for inclusion.**

end of the growing months when all the food was harvested, the remaining pieces were recycled and repurposed.

More than 50 different types of crops were grown, with foods from multiple cultures chosen, including chards, onions, mustard greens, carrots, Huauzontle, and Pigeon Peas. At its completion, Farm for the City produced more than 1,250 pounds of produce, which included more than 250 pounds of lettuce and 53 different plant varieties. The produce was prepared by chefs to provide for more than 8,600 unhoused, at-risk, and low-income individuals.

This installation is one example of urban agriculture in visible and unique practices. Though temporary, the visibility of this can have long-term success when people create policies and practices that support food security for people who need it. Access to healthy food, no matter an individual's income, location, or health, is part of a necessary stable base needed for healthy growth and success. Only when communities develop resources to acknowledge the basic needs and rights – including food, water, and community gathering culture – can they say that they are addressing equity and inclusion in an intentional and conscious approach for short-term and long-term growth.

2.9 Interior Agricultural Green Design

Interior sites are temperature-stable spaces that can be utilized for growing agriculture for communities. The design in Image 2.28 uses an underutilized building space altered with light-emitting diode (LED) plant lights, and with

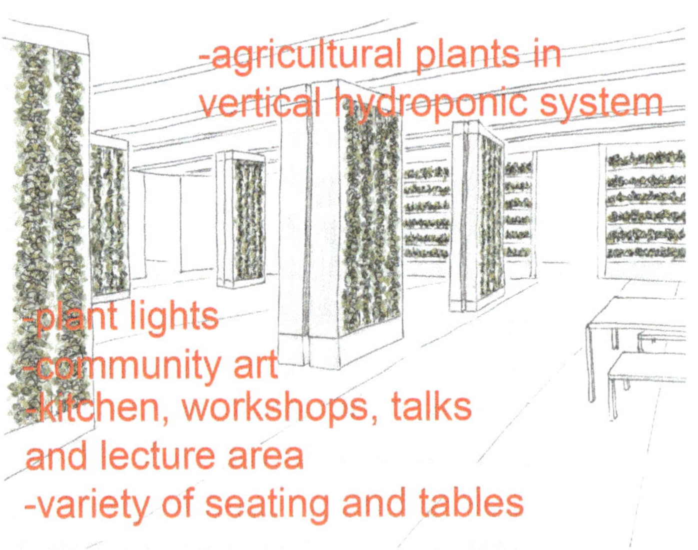

Image 2.27 Underutilized interior spaces are opportunities for year-round food growth, as well as areas for learning and teaching about food access, food preparation, green design, and hydroponics.

the use of captured rainwater from rooftops and greywater, communities can grow healthy food as a local and regional resource for people who want or need it.

Creating a community space includes creating a space for growing, education, and sharing. This community space requires a kitchen and multiple styles of seating and tables to be effective for people to use the space well. The urban community agricultural site is accessible for the people of the community to have access to growing food for the people who want and need it within their community, particularly the underserved and marginalized populations.

Interior hydroponic site designs have the advantage of being able to be grown on a schedule based on the growth cycle of the plants. Once this is established, the volunteer community as well interns, workers, and others can schedule seeding, maintenance and harvesting of the produce. Classes, workshops, talks, and other events can be held in the space.

Having a green interior space available year-round for the community increases not only food security for the community, but also a place for work, volunteering, and to share time in a healthy space, no matter the conditions outside. Interior agriculture allows people who want and need food to have access to it, as well as creating job opportunities for community members to care for and distribute the food to community members, food pantries, education centers, and other community food-sharing programs for food security.

Further, growing plants from around the world, creating regional and national recipes, developing education for the community, and creating a space where equity and inclusion are celebrated are achievable through a successful interior community agricultural site.

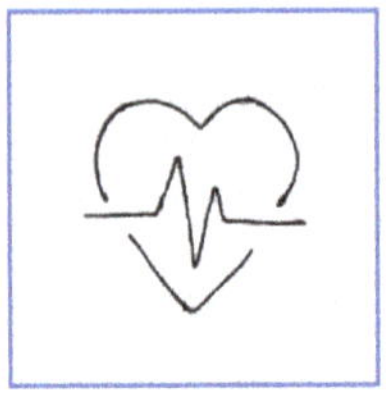

Health – The addition of interior hydroponics for agriculture creates access to year-round healthy local food. This design improves the interior air quality for easier breathing as well as emotional health. The plants selected produce food, which is shared throughout the community.

Abilities – Planters of a variety of heights encourage users in wheelchairs to enjoy and care for the urban agriculture. High-contrast signage improves navigation for people with different visual and walking abilities. A variety of seating options and spaces for people with different mobility skills are available.

Sexuality –Signage of local businesses welcomes and celebrates people of this community.

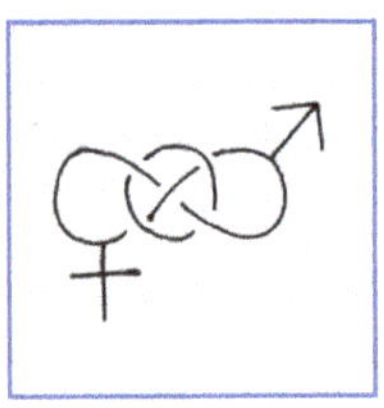

Gender – Creating a source of healthy food that is close by for womxn who work at home is an inclusive choice for their time and health.

Age – Textured walking surfaces throughout the space are for people to safely enjoy the community space. Plantings at a variety of heights and designs further encourage a variety of people of different ages to enjoy the green warehouse. A variety of seating options and spaces for people with different mobility skills are available.

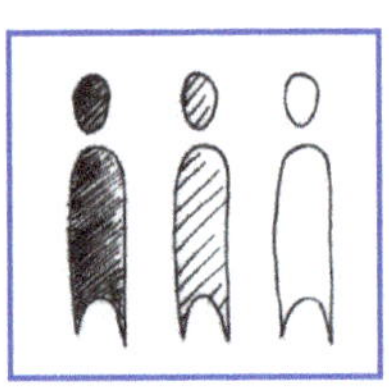

Race – The green warehouse with food grown hydroponically addresses food deserts, which have been created predominantly in communities of people of low income as well as people who are Black, Brown, and underserved. Designs, signs, and street names celebrate people who are Black, Brown, and from underserved communities.

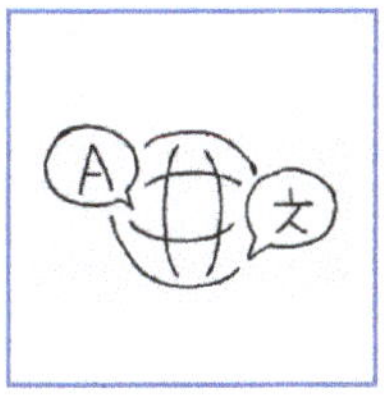

Language – Signs that use multiple local, regional, and international languages, as well as icons and images without words for those learning language, are used throughout the street at pedestrian heights.

Income – The green warehouse addresses food deserts, which have been created predominantly in communities of people of low income as well as people who are Black, Brown, and underserved. The local food is available to eat to support people with low incomes in the community who want or need healthy food. The care for the median can also provide jobs to people who want temporary or long-term jobs.

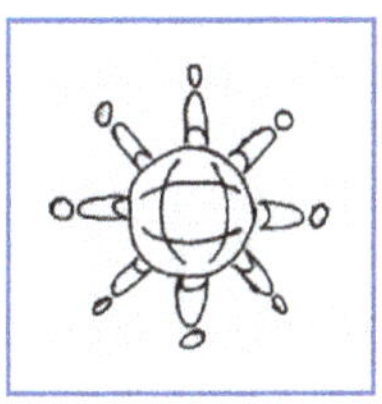

Culture – Cultivating food crops celebrates people of different local, regional, national, and international communities.

Empty buildings such as warehouses are underutilized interior spaces that can be used for community agriculture. Creating an interior agricultural site allows for year-round use in climates that have a colder winter.

Image 2.28 **Interior hydroponic garden for agriculture in a warehouse space.**

Creating an interior hydroponic garden for year-round food growth serves the community for local food security. Food security allows people to have their basic need of food met with a healthy source. Further, this space can also have an educational kitchen for teaching culinary skills, offering food certificates towards job opportunities, and encouraging people of the community to work towards food health for its residents.

Image 2.29 **The indoor hydroponic system works with an efficient design to grow food on a routine schedule.**

2.10 Hydroponic Solutions for Food Growth

Hydroponics systems are those designed for growing plants without the use of soil. Designs for these systems can be primarily horizontal, vertical, or a combination of both. There is an area in the structure in which the seeds are added, which also holds moisture around the seed through a sponge piece.

Water runs through the system for the seed to germinate and grow to mature heights and be harvested. The water is pumped through to the top of the system and then gravity fed to run down through to the bottom. There is a water tank that holds the water. The water in the water tank is routinely

Images 2.30–2.31 **With the known schedule of the food growth, there is a known quantity of food that is produced on a known schedule. This enables companies to schedule harvesting the crop and re-seeding efficiently.**

Image 2.32 **LED lights can change colors for sprouting, leafing, and growth to be most effective at different stages of the plant growth.**

tested for its pH level, altered with base alkaline or acid for a suitable pH. Nutrients for the plants are also added to the water for the plants to use as plant food while they grow in the system.

LED plant lights are set on a timer for the most effective color range for plant growth. Blue light helps plants produce chlorophyll, which is needed for growth. Blue light also encourages vegetative growth, such as developing roots and leaf thickness, and helps young plants germinate. Red light helps

Image 2.33 **Hydroponic systems can be vertical, horizontal, or both in the same design. This allows designs that are best suited for the space they are in, as well as the needs of the growers.**

plants produce flowers, fruit, and stems. Red light also regulates plant growth, flowering, germination, and dormancy. The combined use of both red and blue lights is effective for plant growth.

The lights and irrigation are on a timer. With the known time for plant growth from seed to harvesting, scheduling for the plants produced can be known ahead of time. There are plants that can be grown hydroponically that are already rooted, as well as plants that do not produce food. Cannabis plants have grown well in hydroponic systems for decades, as well as tomatoes and ferns, as some examples. This chapter focuses on food security and therefore understanding how to grow food in hydroponic systems is illustrated in these examples.

Using hydroponic systems for a variety of plants, both edible and non-edible, is also beneficial for a diverse crop. Spending time in green spaces, caring for gardens, and sharing plants helps improve physical and emotional health, so using hydroponic systems for food security – as well as other plants – is an approach toward equity and inclusion.

2.11 Moving Forward with Healthy Food Access

Food justice for equity and inclusion requires interior and exterior approaches, small-scale and large-scale design interventions on public and

Image 2.34 **Adding public community spaces, as well as semi-private spaces, for growing food creates shared practices for health and celebration for not only individuals but community and furthering cultural equity.**

private sites, and cooperation of resources among businesses, residents, and governing agencies. This cooperation leads to food security – a basic right to live a healthy life. With current technologies, design practices, available spaces, and emerging fields available, food security becomes easier to address for the short-term and long-term health of society. Food security and health are directly linked. With green design, connecting the link of available food becomes a reality when communities apply it to their unique sites.

True equity requires creating a community mission in which available healthy food is used not a form of exclusive status, but rather as a basic right to help all members grow. When this is the approach of a community, practices toward this mission become a way in which people can learn and celebrate their neighbors' cultures, traditions, values, and stories to share together.

This chapter and further chapters have green designs, some including interior and exterior plants which produce food, as well as public and private agricultural designs. Health and access to healthy food as a basic right for people – no matter their health, ability, sexuality, gender, age, race, language, income, or culture – are the goals that – with cooperation – are achievable.

References

American Society of Landscape Architects [ASLA]. (2021). Farm for the City, Philadelphia, Pennsylvania, Viridian Landscape Studio. *ASLA 2021 Professional Awards.* Retrieved 20 December 2023, from https://www.asla.org/2021awards/2693.html

Bosch, A., Robles, E. H., & Pool, J. R. (2021). Improving Water Security Helps Reduce the Gender Gap in Mexico. *Cities 4 Forests.* Retrieved 20 December 2023, from https://cities4forests.com/story/improving-water-security-helps-reduce-the-gender-gap-in-mexico-city/

Chapter 3

Sites That Flood

Images 3.1–3.5 **Sites that flood require design interventions that utilize some of the water, as well as areas to move water quickly for residents and business owners to continue to live and grow safely.**

DOI: 10.4324/9781003475293-4

3.1 Introduction

In Chapter 3, green design for addressing the challenges of flooding from small scale through large scale are explored in residential and urban sites. Reducing the effects of flooding events can reduce the cost of repairs to residents, as well as overall communities. These designs are not for addressing catastrophic events, but rather for flooding that happens in typical years in areas near water, as well as areas prone to flooding. Areas prone to flooding can be as a result of being on low land with structures and impermeable surfaces on high land moving water downhill to communities on low ground.

Flooding affects a variety of sites on both large and small scales. The impacts in different communities in the United States are more severe for areas with people of low income and underserved communities. Research in this chapter illustrate that families who are Black, Hispanic, and of underserved communities are more likely to live in high-risk areas due to the lack of information provided about the flood risks of the area (Mittan, 2020).

In order to help reduce flooding in built areas, nature-based solutions are created whereby water can be used in living green designs to reduce the amount of stormwater going into the streets and sewer system. During storm events, slowing down water with permeable surfaces and living designs that can use the stormwater reduces the water flow to low points and drains. Drains in the streets or curbs may not be able to take in the amount of water, or may be clogged due to debris, causing slow or no drainage for extended periods of time.

Green designs can be created on a neighborhood scale with planned living spaces. Further, when each residential lot adds a notable presence of green design to their site, the totality of all the living spaces is beneficial to reduce stormwater. The same intention applies to urban sites for adding green design on both a community scale and in single buildings. The combined effects of living design in areas with fewer permeable surfaces can reduce the damage caused by storms.

Creating policies and practices means that once the risk of flooding is reduced or removed due to green design, the people who live in the newly redesigned areas can stay in their safe communities without raised prices. When designers create spaces for people who are marginalized based on their health, abilities, sexuality, gender, age, race, language, income, or culture, the spaces become inclusive, safer, and encourage a greater use of the area. These goals are important for communities to grow in the short term and the long term.

3.2 Suburban Community Green Design

The effects of flooding are reduced with green design. The suburban community in Image 3.7 has layers of greening designed throughout this area of small through large scale. It is the totality of the multiple interventions with greening and permeable surfaces – all of which use rainwater and greywater for irrigation – that benefits the structures with less damage due to water overflow and flooding.

The large-scale aerial view of the spring scene of the suburban green design space focuses on reducing flooding to address equitable design for its residents. This includes features such as multiple forested areas embedded within the properties with houses; green roofs throughout the buildings; expanded tree wells with gardens on the sidewalk; planted front, side, and back yards; and permeable pavers in driveways and alleys.

Green design throughout the community in small and large interventions reduces the effects of stormwater, even with large amounts of water. This becomes important when understanding the structural effects and financial damage and devastation that stormwater can cause. Reducing the effects – whether by a small, medium, or large amount – all adds up to reducing the financial devastation to individuals, families, and communities.

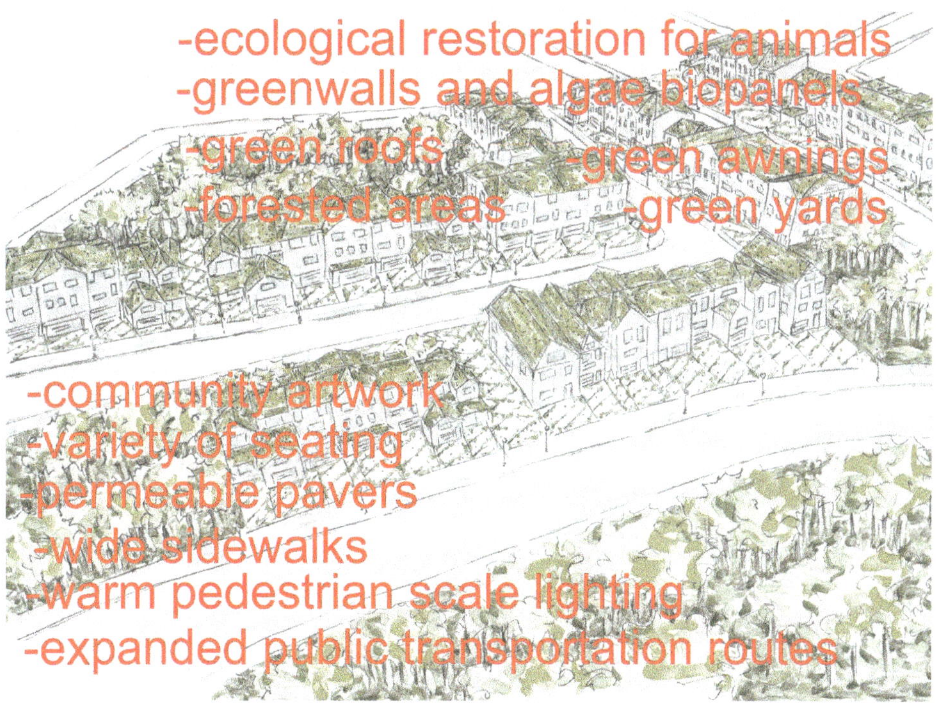

Image 3.6 On a community scale, the layers of green intervention in a residential area are able to use and filter rainwater, as well as reduce the effects of larger storm events.

Further equitable and inclusive design, in addition to addressing issues of flooding, include widening sidewalks for walking comfort and safety; adding a sidewalk zone created for consistent pedestrian-scale warm lighting, trashcans, and a variety of seating styles for rest and other public comforts; adding public transportation routes; and creating murals on sidewalks celebrating culture and diversity.

The green design also adds space for birds and other displaced animals to live, improving the ecology in the community. Additionally, shared playgrounds for the community create invested spaces for families over time. Trees and plants throughout the community space and residential space produce food for people who want or need it. The totality of the green design interventions improves air quality by reducing PM 2.5 (fine particulate matter measuring less than 2.5 micrometers in diameter) and contaminants in the air. Planted spaces reduce the strength of sound waves, which reduces noise pollution.

With colorfully painted sidewalks wide enough for ease of walking for inclusive use of people with different walking abilities, as well as people with children, a greater sense of community is created. This creates an inclusive neighborhood with more ease for family and friends.

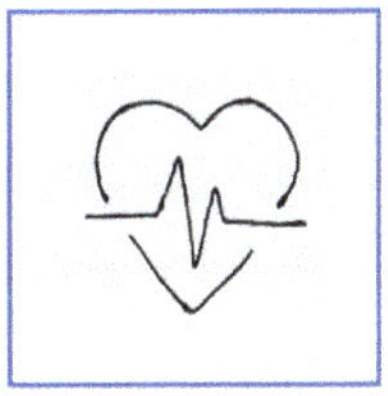

Health – Reducing the effects of stormwater through green design improves people's financial health, emotional health, and physical health. The addition of greening also improves air quality for easier breathing as well as for emotional health. The plants selected produce food, which is shared throughout the community. Street murals use air-purifying paint, which absorbs and neutralizes chemicals, pollutants, and VOCs for improved air quality. Front yard and sidewalk seating create a space for stopping, sitting, or community gathering.

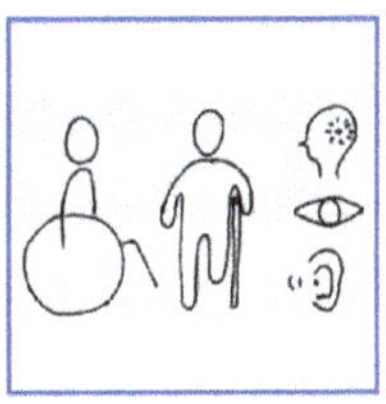

Abilities – A variety of front yard and sidewalk seating options for people with different mobility skills are available for a healthy walk or while running errands. The street murals are colors visible to people with different visual abilities. Curb cuts, detectable rumble strips, and warning pavers at corners; crosswalk melodies for people who are walking and do not walk; and high-contrast signage improve navigation for people with different visual and walking abilities. With wider sidewalks, there are zones for pedestrians to move at their pace.

Sexuality – Consistent luminescent lighting at pedestrian scale along each block, as well as warm lighting, improves safety for the LGBTQIA+ community. Signage of local businesses welcomes and celebrates people of this community.

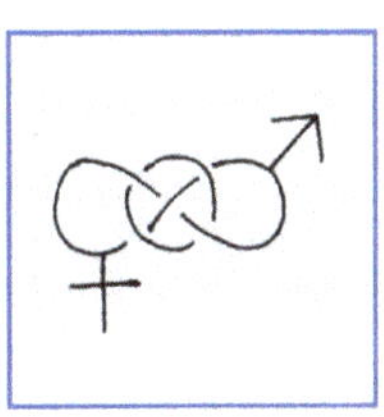

Gender – The center median allows a resting or stopping spaces for caregivers with children and senior adults. Consistent luminescent lighting at pedestrian scale along each block, as well as warm lighting, improves safety for womxn in the community. Wider sidewalks are spaces for strollers and holding children's hands. Public transportation is routed for womxn and people who run errands and take routes with multiple stops. Designs, signs, and street names celebrate womxn.

Age – A variety of front yard and sidewalk seating options for people with children or who are caregivers are available for a healthy walk or while running errands. Consistent luminescent lighting at pedestrian scale along each block, as well as warm lighting, improves safety for womxn in the community. Wider sidewalks are spaces for strollers and holding children's hands. Public transportation is routed for womxn and people who run errands and take routes with multiple stops. Designs, signs, and street names celebrate womxn.

Race – Designs, signs, and street names celebrate people of different local, regional, national, and international communities.

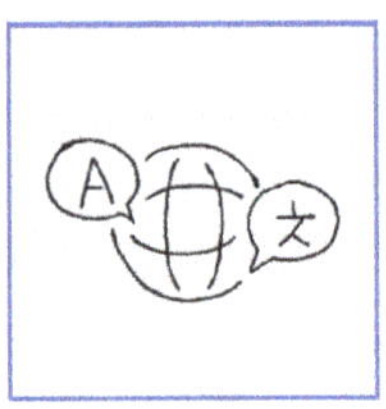

Language – Signs that use multiple local, regional, and international languages, as well as icons and images without words for those learning language, are used throughout the street at pedestrian heights.

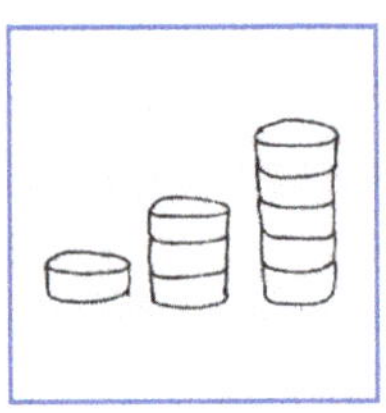

Income – Green designs use rain and stormwater, reducing the structural effects and other costs of storm events. Plants that produce food are included in front yards, sidewalk-expanded tree wells, and forested areas. This approach supports people with low income in the community who want or need healthy food. The care and distribution of these spaces can also provide jobs to people who want temporary or long-term jobs.

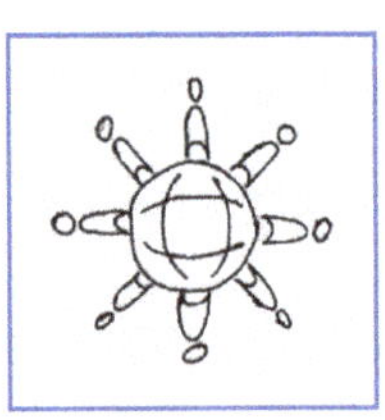

Culture – Food, designs, signs, and street names celebrate people of different local, regional, national, and international communities.

Greening suburban areas from the community to the site scale reduces the effects of storm events for the families who live there, the animal population, and overall community health.

Image 3.7 **Overview of a green suburban community.**

This bird's-eye view of a suburban area utilizes green design to reduce rainwater and stormwater for small and large rain events. Green roofs, permeable pavers in the driveways and paths, forested lots, and open areas allow stormwater to be utilized for irrigation in the planted areas, as well as slowing down and reducing the stormwater before it gets to city sewer systems. This green approach reduces the stress on the municipal and regional sewer systems. The totality of an entire community creating green design can reduce the damage – as well as the cost – of repairs from storm events.

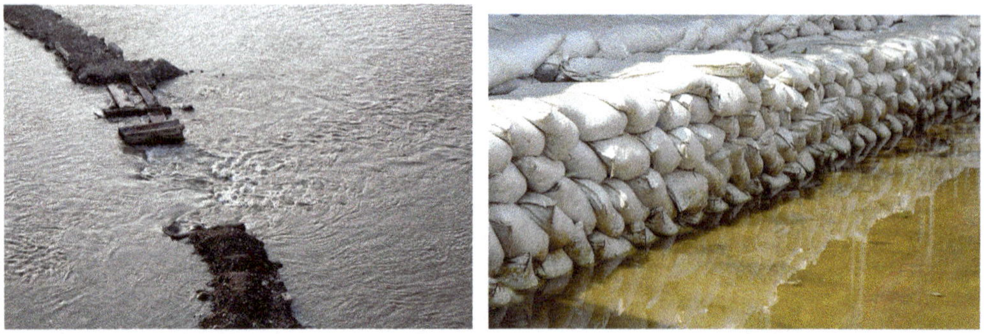

Images 3.8–3.9 The effects of flooding range from smaller private sites through larger public comforts.

3.3 Black and Hispanic People More Likely to Live in High-Risk Flood Zones

A study by Laura Bakkensen, ecological economist, professor at the University of Arizona, explored the issues of different costs of insurance in high-risk flood areas for people who are Black and Hispanic with low incomes (Mittan, 2020). It found that people who are Black and Hispanic with low incomes are more likely to live in areas with high risk of flooding than people who are white and Asian. Further, the flood insurance program administered by the federal administration disproportionately costs more to people who are Black and Mexican of low income.

The study findings came during 2020, a year of record-breaking storms across the Atlantic seaboard in the United States. The study focused on sites in southern Florida prone to flooding, the property value and sales, and the ethnicity and income of the homebuyers. The question became clear, said

Images 3.10–3.11 Flooding disproportionately affects people who are Black and Mexican with lower income (Mittan, 2020).

Images 3.12–3.13 The damage from flooding affects the short-term and long-term financial, physical, and emotional health of individuals, families, and the larger community.

Bakkensen: "How much would somebody be willing or able to pay to avoid living in a high flood-risk zone?" The results were disproportionate, with higher costs for people who are Black or Mexican with lower income, and lower rates for people who are white or Asian.

"We see a lot of news reports where sometimes poorer or minority-rich communities may feel greater losses," Bakkensen said. There are multiple causes of these different costs. One is the access or lack of access to information about flood risk. Another is the use of old flood maps for low and high risks, as they are unreliable as related to the current conditions. Updating these maps adds value to the properties – which is beneficial for people. The added values of the properties varied, with lower values for people who are Black and Mexican in contrast to people who are white or Asian.

In a related study in 2023 about flood zones, it was found that people who are Black, Hispanic, and Native Americans are more likely to live in

Images 3.14, 3.15 People who are Black, Hispanic, and Native American are more likely to live in flood zones with sub-par levees, which do not protect their homes (Mittan, 2020).

flood zones and with sub-par levees, which do not effectively protect their homes (Vahedifard et al., 2023). Minority groups are more likely to live in low-lying areas with poor drainage systems, exposing their homes to flooding.

"The evolving risk of flooding is shown to disproportionately affect historically underserved and socially vulnerable communities," said Farshid Vahedifard, the lead author of the study. In 2022, the Biden/Harris administration proposed $1.2 billion in federal funds for infrastructure to help minority communities, such as those living near failing levees.

3.4 Suburban Neighborhood Green Design

In looking at a closer scale of a neighborhood, the layers of green design are used to reduce the effects of flooding in small and large interventions.

The design in Image 3.17 uses green roofs, green walls, and permeable pavers planted throughout the suburban neighborhood. Balcony plantings, plantings on short walls, and street trees, along with shrubs, tall wavy grasses, and other plants, are designed throughout the neighborhood. The permeable pavers are designed with plantings that allow for water to drain, as well as for people to walk or use a wheelchair, walker, or stroller.

Stormwater is the primary source of irrigation, with greywater from the residential buildings being the secondary source. Water recycling helps lower the cost of the maintenance of the site. Creating green, open, and permeable space throughout the neighborhood reduces stormwater damage and the cost of

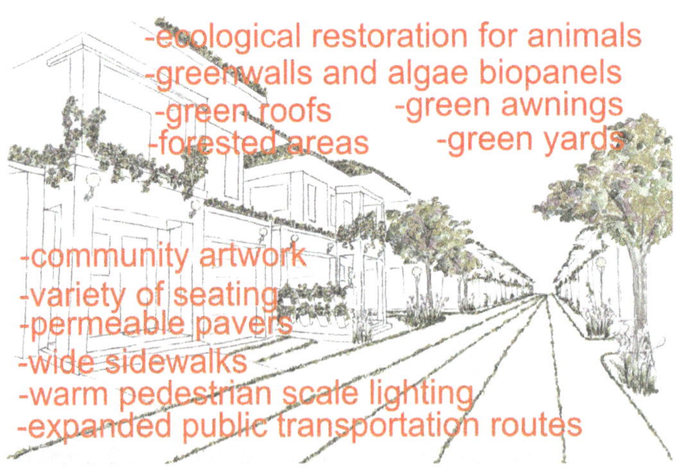

Image 3.16 **On a more focused street scale, the layers of green permeable surfaces use rainwater and stormwater for growth, reducing the amount and pace of the water that goes to the overworked city systems for cleaning. Other layers of green design also reduce stormwater runoff.**

repairs. The totality of each resident and shared space having applied these green practices adds to the total reduction of damage and repairs during a storm event.

Colorfully painted sidewalks that are wide for ease of walking and are inclusive for use by people with different walking abilities, as well as people with children, this creates an inviting community for a variety of people. Lighting is warm in color and consistent in spacing, adding safety throughout the neighborhood for walking, biking, running, or strolling. Seats are placed at intersections as resting points for users.

The green design adds scented plants for physical and emotional health, as well as a space for birds and other displaced animals to live. The green design improves air quality and reduces noise pollution. The effects of the layers of green design on flooding are also beneficial for inclusion in design for a diverse range of people based on their health, abilities, sexuality, gender, age, race, language, income, and culture.

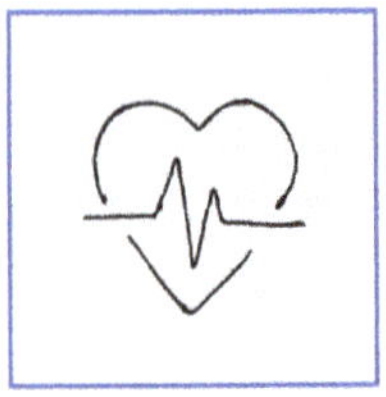

Health – Reducing the effects of stormwater through green design improves people's financial health, emotional health, and physical health. The addition of greening also improves air quality for easier breathing as well as for emotional health. The plants selected produce food, which is shared throughout the community. Street murals use air-purifying paint, which absorbs and neutralizes chemicals, pollutants, and VOCs for improved air quality. Front yard and sidewalk seating create a space for stopping, sitting, or community gathering.

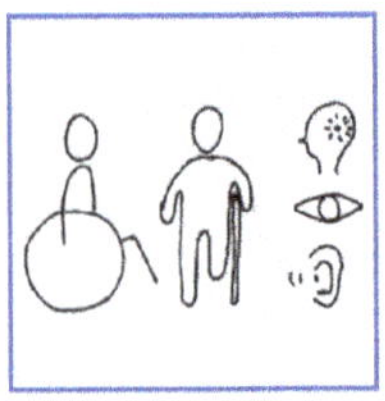

Abilities – A variety of front yard and sidewalk seating options for people with different mobility skills are available for a healthy walking or while running errands. The street murals are of colors visible to people with different visual abilities. Curb cuts, detectable rumble strips and warning pavers at corners; crosswalk melodies for walking and do not walk; and high-contrast signage improve navigation for people with different visual and walking abilities. With wider sidewalks, there are zones for pedestrians to move at their pace.

Sexuality – Consistent luminescent lighting at pedestrian scale along each block, as well as warm lighting, improves safety for the LGBTQIA+ community. Signage of local businesses welcomes and celebrates people of this community.

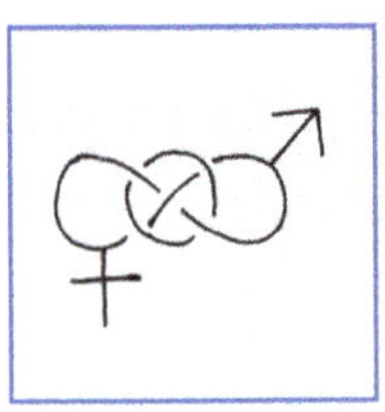

Gender – A variety of front yard and sidewalk seating options for people with children or who are caregivers are available for healthy walking or while running errands. Consistent luminescent lighting at pedestrian scale along each block, as well as warm lighting, improves safety for womxn in the community. Wider sidewalks are spaces for strollers and holding children's hands. Public transportation is routed for womxn and people who run errands and take routes with multiple stops.

Age – A variety of front yard and sidewalk designs and seating options are available for young children and mature adults as they are walking or running errands. Curb cuts and safe crosswalks are at every street corner for people to safely cross streets at multiple locations on their route. This is important for people who walk throughout their day while running errands with children and mature people.

Race – Designs, signs, and street names celebrate people of different local, regional, national, and international communities.

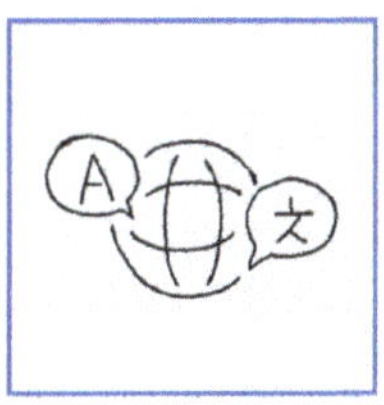

Language – Signs that use multiple local, regional, and international languages, as well as icons and images without words for those learning language, are used throughout the street at pedestrian heights.

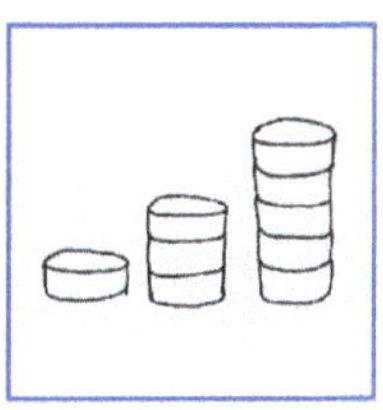

Income – Green designs use rain and stormwater, reducing the structural effects and costs of storm events. Plants that produce food are included in front yards, sidewalk-expanded tree wells, and forested areas. This approach supports people with low income in the community who want or need healthy food. The care and distribution for spaces can also provide jobs to people who want temporary or long-term jobs.

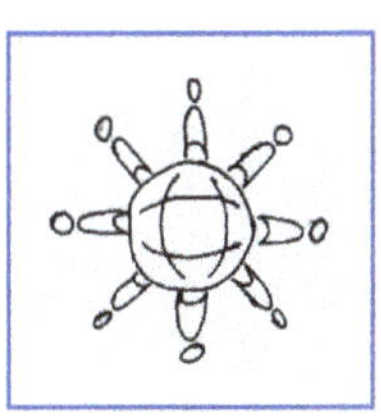

Culture – Food, designs, signs, and street names celebrate people of different local, regional, national, and international communities.

On a street scale, greening suburban communities address reducing the effects of storm events for the families who live there, the animal population, and the overall community health.

Image 3.17 **Greening design for a suburban street.**

On a more focused scale, this image illustrates green design for rainwater and stormwater on a neighborhood site. Green roofs, pathways, expanded tree wells, and yard greening use rainwater and stormwater. This reduces the quantity and speed of the rainwater and stormwater going to the city sewers. The reduced stormwater can help overworked municipal water plants, as well as combined sewer systems, to be more efficient.

Images 3.18–3.19 **The lack of permeable surfaces in public spaces is damaging communities when there are high-intensity rainfalls.**

3.5 The Reasons for Jakarta's Frequent Flooding and How Nature-Based Solutions (NBS) Can Help Reduce the Risk

There are multiple key factors that caused the 2021 floods in Jakarta (Wihanesta et al., 2021). The first was the high-intensity rainfalls as a result of the climate crisis. Secondly, the land cover had shifted drastically with settled areas growing 47.4% from 2000–2019. The settled lands replaced green open spaces, which by 2019 were reduced to 9.8% of Jakarta. The land cover shift causes higher volume of runoff water, as well as high erosion rates in the streams. Finally, there is land subsidence – the gradual sinking or caving of land – with a range of 12–25 centimeters per year throughout different areas of the Indonesian city. Land subsidence has grown as a result of increased

Images 3.20–3.21 **With climate changes, which have shifted over decades, the quantity and intensity of high-intensity rainfalls have increased (Wihanesta et al., 2021).**

Image 3.22–3.23 **Creating safe, healthy, and affordable living sites for people is a basic inclusive approach for designing spaces.**

building loads and excessive ground water extraction – both of which have accelerated the expansion of sinking land.

The current practice in Jakarta is to focus on physical structure to address the lack of drainage of flood water. However, this is a limited solution that has been shown to fail, such as in the example of the flood wall structurally failing in New Orleans, Louisiana during Hurricane Katrina.

Image 3.24 **This design is used for moving stormwater safely and quickly through an area with multiple locations for water to drain.**

Nature-based restoration is the use of green infrastructure and can accompany structural solutions. These solutions range from small scale to large scale. This includes water infiltration holes in the soil, green roofs, and rainwater collection tanks on a small scale. On a large scale, nature-based solutions (NBS) include multi-functional green open spaces such as areas to catch water, bioswales, and permeable pavers.

Not only in Jakarta is this a relevant practice; it also applies to other developed areas prone to flooding. The intervention of greening in small-scale and large-scale private and public spaces, on structures as well as areas with soil, all add to spaces where water can infiltrate the ground rather than running to low land. The use of layered greening, along with short-term and long-term structures, designs, plans, and policies for communities to address flooding is a also required for the long-term stability, safety, and growth of people and communities.

3.6 Street and House Green Design

In looking at a house scale, Image 3.26 illustrates the layers of green design applied with green roofs, green awnings, green windowsills, green vertical surfaces, and planted front yards with planters or open surfaces. Expanded tree wells include shrubs, tall grasses, and other plants. Permeable pavers are in front yards and driveways and are designed with plantings that allow for water to drain, as well as for people to walk or use a wheelchair, walker, or stroller.

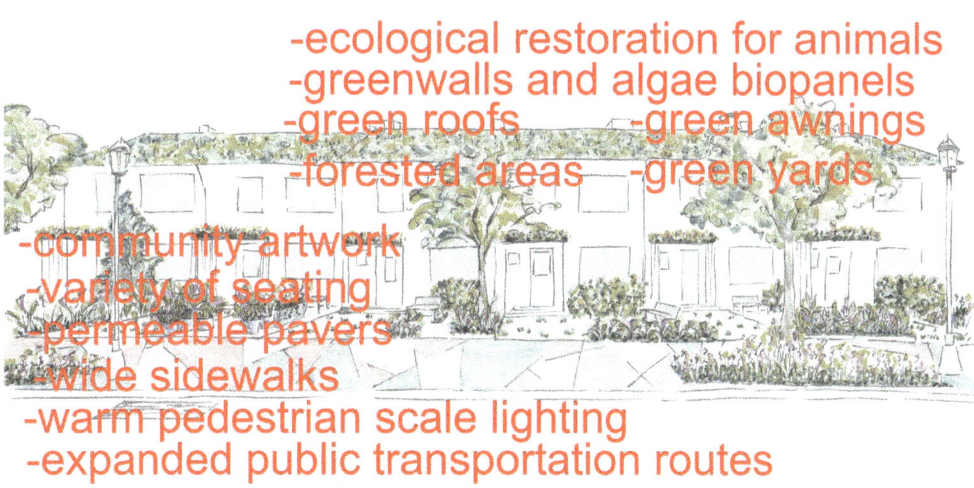

Image 3.25 Green design at a focused house scale, when added together, benefits a larger scale of neighborhoods and communities to use stormwater during rain events, reducing flooding.

Stormwater is the primary source of irrigation, with greywater from the residential buildings being the secondary source. Water recycling helps lower the cost of the maintenance of the site. Creating as much green, open, and permeable space throughout the neighborhood as possible reduces stormwater damage and the cost of repairs. The totality of each residential and shared space having applied these green practices adds to the total reduction of damage and repairs during a storm event.

Front yards have seating to create a space for sharing time outside with the community. Seating is also added in the sidewalk area in order for users to rest throughout a neighborhood walk or while waiting for public transportation. The sidewalk is expanded for ease of use for people with strollers, people with children, people in wheelchairs, people with walkers, and other users to share the space.

Colorfully painted sidewalks that are wide for ease of walking are inclusive for the use of people with different walking abilities, as well as people with children. This creates an inviting community for a variety of people. Lighting is warm in color and consistent, adding safety throughout the neighborhood for walking, biking, running, or strolling.

The green design adds flowering, edible, and scented plants for physical and emotional health, as well as a space for birds and other displaced animals to live. The green design improves air quality and reduces noise pollution. The effects of the layers of green design for flooding are also beneficial for a diverse range of people based on their health, abilities, sexuality, gender, age, race, language, income, and culture.

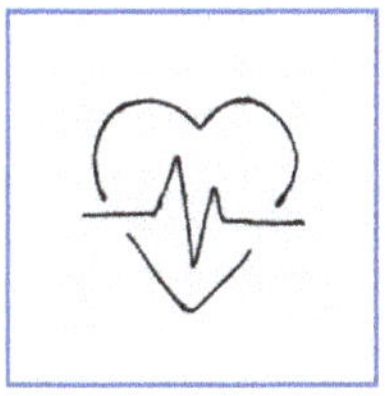

Health – Reducing the effects of stormwater through green design improves people's financial health, emotional health, and physical health. The addition of greening also improves air quality for easier breathing as well as for emotional health. The plants selected produce food, which is shared throughout the community. Street murals use air-purifying paint, which absorbs and neutralizes chemicals, pollutants, and VOCs for improved air quality. Sidewalk seating creates a space for stopping, sitting, or community gathering.

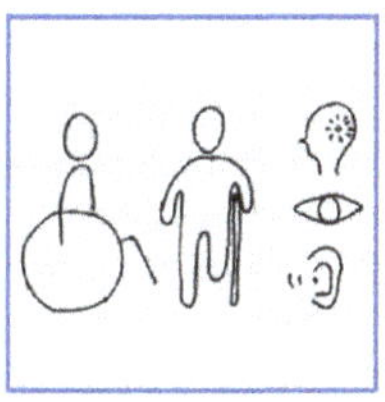

Abilities – A variety of front yard and sidewalk seating options for people with different mobility skills are available for a healthy walking or while running errands. The street murals are of colors visible to people with different visual abilities. Curb cuts, detectable rumble strips, warning pavers at corners, crosswalk melodies for walking and do not walk, and high-contrast signage improve navigation for people with different visual and walking abilities. With wider sidewalks, there are zones for pedestrians to move at their pace.

Sexuality – Consistent luminescent lighting at pedestrian scale along each block, as well as warm lighting, improves safety for the LGBTQIA+ community. Signage of local businesses welcomes and celebrates people of this community.

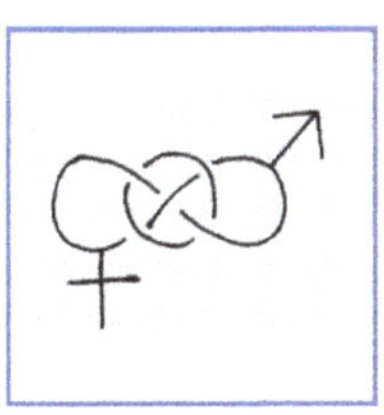

Gender – A variety of front yard and sidewalk seating options for people with children or who are caregivers are available for healthy walking or while running errands. Consistent luminescent lighting at pedestrian scale along each block, as well as warm lighting, improves safety for womxn in the community. Wide sidewalks are spaces for strollers and holding children's hands. Public transportation is routed for womxn and people who run errands and take routes with multiple stops.

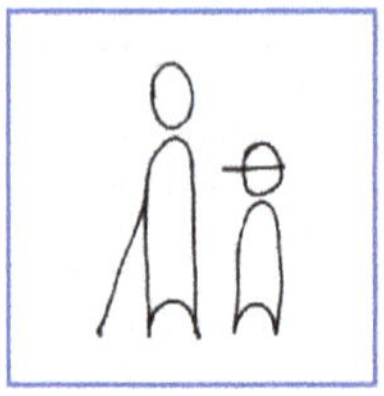

Age – A variety of front yard and sidewalk designs and seating options are available for young children and mature adults as they are walking or running errands. Curb cuts and safe crosswalks are at every street corner for people to safely cross streets at multiple locations on their route. This is important for people who walk throughout their day while running errands with children and mature people.

Race – Designs, signs, and street names celebrate people of different local, regional, national, and international communities.

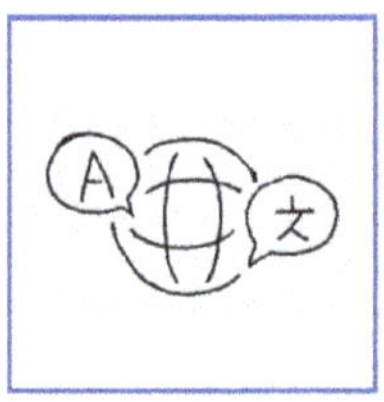

Language – Signs that use multiple local, regional, and international languages, as well as icons and images without words for those learning language, are used throughout the street at pedestrian heights.

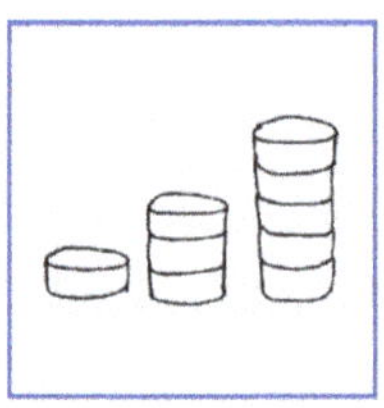

Income – Green designs use rain and stormwater, reducing the effects and costs of storm events. Plants that produce food are included in front yards, sidewalk-expanded tree wells, and forested areas. This approach supports people with low income in the community who want or need healthy food. The care and distribution of these spaces can also provide jobs to people who want temporary or long-term jobs.

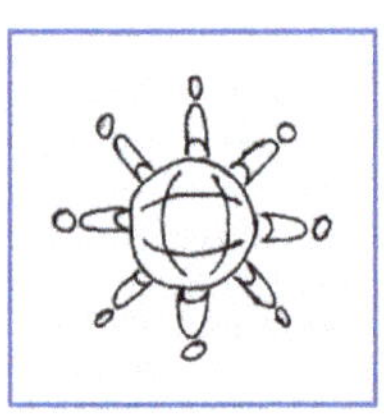

Culture – Food, designs, signs, and street names celebrate people of different local, regional, national, and international communities.

On a house scale, greening suburban communities addresses reducing the effects of storm events for the families who live there, the animal population, and the overall community health.

Image 3.26 **House-by-house scale in suburban neighborhood.**

In focusing from a large overview to a house scale, the opportunities for rainwater and stormwater reduction can both be transformative and provide an inclusive approach to design. This image illustrates stormwater reduction through sloped green roofs; green planted front awnings; green front yard designs, both with pots as well as planted into the ground; and permeable pavers and walkways with living ground cover and expanded tree wells on the sidewalks. Sidewalks are also wider to allow for strollers, people with different mobilities, and benches and street lighting.

Images 3.27 **In 2019, there were 103 days of the year on which the riverfront was flooded in Davenport, Iowa.**

3.7 Mississippi River Flood Resiliency Plan, Creating a More Flood-Resilient, Accessible, and Beautiful Riverfront for Davenport, Iowa

The frequency of flooding in and along the downtown waterfront in Davenport, Iowa, has increased significantly in the last 20 years, creating short-term and long-term needs for addressing flooding issues (Sasaki Associates, n.d.). In 2019, there was record flooding. in which 103 days with

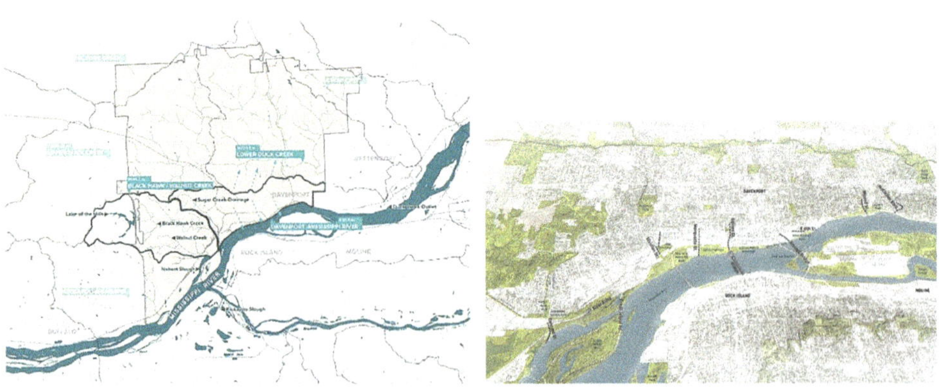

Images 3.28–3.29 **Sasaki and Associates created a Mississippi River flood resiliency plan as a design solution for Davenport, Iowa.**

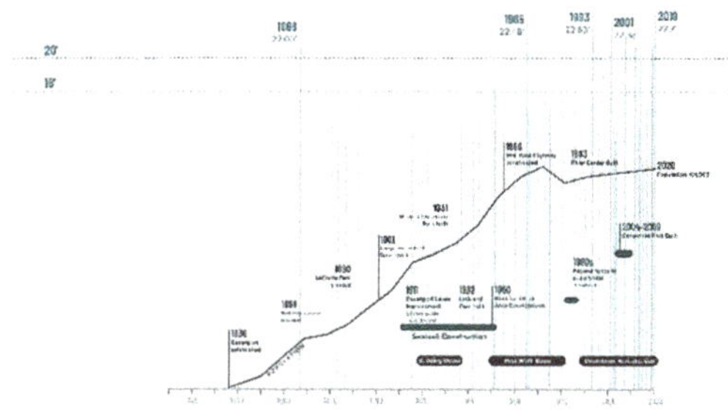

Images 3.30–3.31 The design solution was created as a result of talking with the community and listening to their questions and wants for their community.

of the city riverfront flooded on 103 days. Sasaki and Associates, with HR Green and Shive Hattery, Davenport Public Works, and the community, worked together to create a three-phase plan to address the nine miles of waterfront in the downtown area.

Meetings with the community – including talking with organizations focused on diversity, equity, and inclusion; community leaders; high school students; and local media – were integrated into the planning process with a goal of reducing long-term delays, reducing the drain on city resources and lack of access to lots of public spaces caused by the flooding. The groups also studied the riverfront neighborhoods in the context of the watersheds and creeks that drain into the Mississippi River.

The plan to address flooding also incorporates a line of protection and other public benefits in the area, including an elevated open space and trail connectors with both above-ground and below-ground structural changes. Structural solutions include riverfront berms, floodgates, and infrastructure for both short-term and long-term flooding, as well as non-structural approaches such as land use changes.

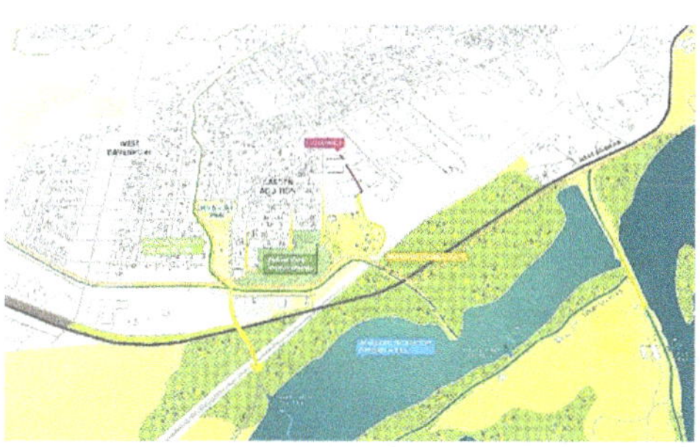

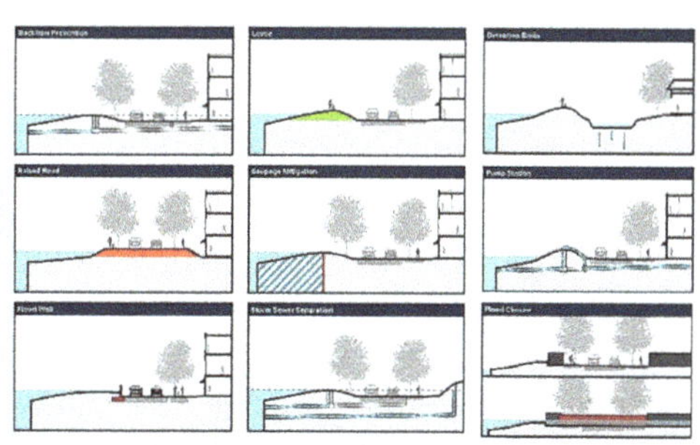

Images 3.32–3.33 The design solution created by Sasaki and Associates includes redesigning both public and private sites to address diversity, equity, and inclusion for the community to grow and thrive.

This solution was developed by working with the community and listening to the effects of flooding, including loss of property, loss of work, loss of access to be able to get to work, loss of access to roadways, loss of access to public spaces, extended commuting times, loss of public transportation routes, loss of playground areas, and other effects. Redesigning the area, while making sure these residents are not priced out of their neighborhood once the flood plan is complete, requires not only design but also policies and regulations that require funding put toward the area to prevent gentrification, including green gentrification.

Green gentrification is when green design is added to a low-income or underserved neighborhood for equity and instead forces the same underserved or low-income people to be displaced as people with high income move in and raise the cost of living in the area. This forces the original residents out of the area.

3.8 Urban Community Green Site Design

Looking at a community urban scale in Image 3.35 is an opportunity to be creative, with a myriad of green design interventions to reduce the effects of stormwater and create spaces that are inclusive, as well as using green design toward equity.

A mixed-use urban area has green design woven in layers throughout the space in order to reduce the effects of stormwater, as well as adding green space to improve the physical and emotional health of the residents, employees, and visitors.

Image 3.34 **In mixed-use urban areas, green design interventions throughout a neighborhood add together to benefit the community. The rainwater is used throughout the area for the health and growth of the planted material while also reducing overflow from sewer systems after large, fast quantities of rain.**

Gren design is applied to surfaces such as vertical green walls, building corners, green roofs, balconies, windowsills, awnings, train stops, and bus stops. Expanded tree wells add more layers of green design. The expanded tree wells include trees, shrubs, tall grasses, and other plants, as well as seating for people to be in the community space or while waiting for public transportation.

Stormwater is the primary source of irrigation, with greywater from the buildings being the secondary source. Water recycling helps lower the cost of the maintenance of the site. Creating as much green, open, and permeable space throughout the neighborhood as possible reduces stormwater damage and cost of repairs. The totality of each area that can use stormwater adds to the total reduction of damage and repairs during a storm event.

Lighting is warm in color and consistent, adding safety throughout the neighborhood for walking, biking, running, or strolling. In urban areas where there is activity throughout the day and night, creating safety for pedestrians with warm well-lit routes is an approach to inclusion for the community.

The green design adds flowering, edible, and scented plants for physical and emotional health, improving air quality, reducing sound pollution, and creating views of green design from windows. Additionally, the green design creates living spaces for birds and other animals that have been displaced due to developing areas. The effects of the layers of green design for flooding are also beneficial to a diverse range of people based on their health, abilities, sexuality, gender, age, race, language, income, and culture.

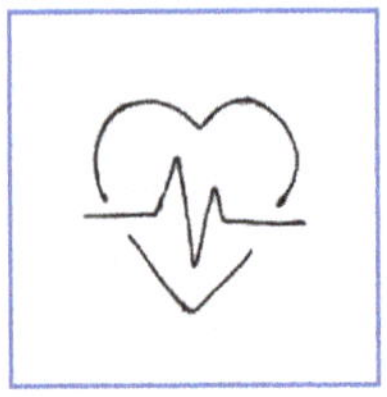

Health – Reducing the effects of stormwater through green design improves people's financial health, emotional health, and physical health. The addition of greening to a denser urban population also improves air quality for easier breathing as well as for emotional health. The plants selected produce food, which is shared throughout the community. Street murals use air-purifying paint, which absorbs and neutralizes chemicals, pollutants, and VOCs for improved air quality. Front yard and sidewalk seating create a space for stopping, sitting, or community gathering.

Abilities – A variety of sidewalk seating options for people with different mobility skills are available for healthy walking or while running errands. The murals are of colors visible to people with different visual abilities. Curb cuts, detectable rumble strips and warning pavers at corners, crosswalk melodies for walking and do not walk, and high-contrast signage improve navigation for people with different visual and walking abilities. With wider sidewalks, there are zones for pedestrians to move at their pace.

Sexuality – Consistent luminescent lighting at pedestrian scale along each block, as well as warm lighting, improves safety for the LGBTQIA+ community. Signage of local businesses welcomes and celebrates people of this community.

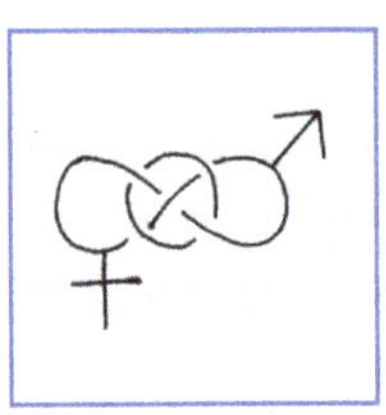

Gender – A variety of sidewalk seating options for people with children or who are caregivers are available for healthy walking or while running errands. Consistent luminescent lighting at pedestrian scale along each block, as well as warm lighting, improves safety for womxn in the community. Wider sidewalks are spaces for strollers and holding children's hands. Public transportation is routed for womxn and people who run errands and take routes with multiple stops. Designs, signs, and street names celebrate womxn.

Age – A variety of sidewalk designs and seating options are available for young children and mature adults as they are walking or running errands. Curb cuts and safe crosswalks are at every street corner for people to safely cross streets at multiple locations on their route. This is important for people who walk throughout their day while running errands with children and mature people.

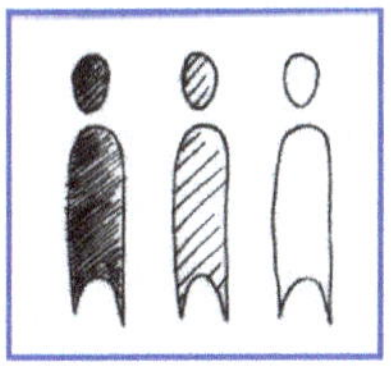

Race – Designs, signs, and street names celebrate people of different local, regional, national, and international communities.

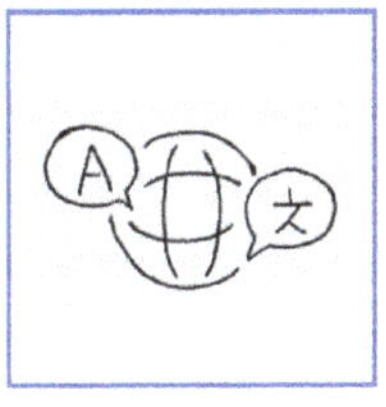

Language – Signs that use multiple local, regional, and international languages, as well as icons and images without words for those learning language, are used throughout the street at pedestrian heights.

Income – Green designs use rain and stormwater, reducing the effects and costs of storm events. Plants that produce food are planted throughout the community including green roofs, green balconies, green walls, expanded tree wells, and vertical greening. This approach supports people with low income in the community who want or need healthy food. The care and distribution of these spaces can also provide jobs to people who want temporary or long-term jobs.

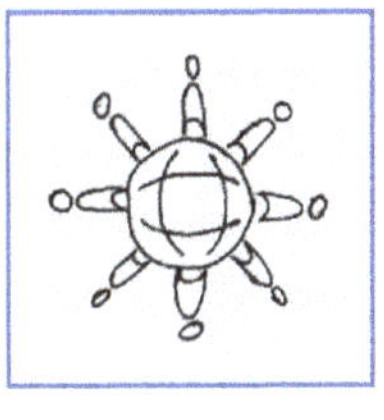

Culture – Food, designs, signs, and street names celebrate people of different local, regional, national, and international communities.

On a community scale, in an urban area, green design reduces the effects of rainwater and stormwater on the structures, city sewer, or combined sewer systems. The vertical and horizontal green designs utilize greywater, rainwater, and stormwater for irrigation with permeable surfaces.

Image 3.35 **Urban community with green design from the eye-level pedestrian view allows people to see the site as they would experience it.**

The green design reduces rainwater and stormwater using green infrastructure and interventions throughout the community, both vertically and horizontally. Green-covered train stations, green-covered bus stops, green roofs, multiple-story green walls, balcony plantings, windowsill planters, and expanded tree wells add living spaces that utilize greywater, rainwater, and stormwater throughout the urban fabric. Green design uses rainwater and stormwater, and slows down the water going to the city sewer or combined sewer system.

Image 3.36 **Greening along roadways includes vertical surfaces and plants that can live and grow on bricks and stones even with the presence of car exhaust. There are plants that can do this well.**

3.9 Nature-Based Solutions (NBS)

There are multiple companies that develop green solutions for coastal flooding and erosion, river flooding and erosion, and urban stormwater flooding. Their goals include using green design to create healthier spaces, improve social ties, and grow healthier communities and stronger economies.

There are different ways in which companies and designs use NBS to reduce the effects of flooding on sites. Some of these approaches are

Image 3.37 **Hydroseeding is used to control erosion by seeding spaces, including those with severe slopes, with a seed mixed with mulch, fertilizer, water, and other materials. This slurry mix is applied with high pressure and grows quickly on site.**

Image 3.38 **Along waterfronts, plants that can tolerate the changing water levels, winds, and weather are selected for effective designs.**

illustrated in examples in this chapter. These green approaches include the use of planted designs, permeable surfaces, areas that bring the water through soil and similar materials, and areas that slow down water in order to allow the ground or permeable surfaces time to absorb the water.

This is effective to slow down the rate of stormwater before it goes to roadways and other non-permeable areas that are unable to filter the water easily and safely. Permeable surfaces and areas; planted vertical and horizontal areas; and planted or open fields, ponds, and lakes are able to absorb and use the stormwater so it does not rush to damage areas.

Image 3.39 **Planting for erosion control includes choosing plants with deep-growing roots that can weave into the soil in order to create a strong structure in order prevent soil slipping during rain events.**

Adding these areas of green design requires green infrastructure and interventions in public and private sites. Making sure planted fields stay as planted fields, rather than being developed into buildings, requires a committed administration to understand the needs for green areas in reducing the effects of flooding.

Local, state, and federal cooperation is required to make NBS effective. The cooperation of individuals, families, communities, and businesses with local, state, and federal agencies furthers the change toward effective solutions. The devastation of flooding can be reduced with a shared understanding of short-term and long-term green interventions and infrastructure.

Storm events will continue, and preparing for them by retrofitting spaces with NBS is an approach that not only addresses flooding; it also has numerous community benefits when there is no stormwater event. These benefits include improved air quality, healthy food, improved physical and emotional health, and improved quality of life. It is the totality of the benefits that makes NBS an approach that is illustrated for equity and inclusion.

3.10 Urban Neighborhood Green Design

A mixed-use urban area in Image 3.41 has green design woven in layers throughout the space in order to reduce the effects of stormwater, as well as added green space to improve the physical and emotional health of residents, employees, and visitors.

Image 3.40 On a more focused street scale of a mixed-use urban setting, green design intervention add together to benefit the community. Vertical and horizontal greening redefines the scene to physically reduce stormwater damage and financially protect people from the cost of damaged property.

Green design is applied to surfaces such as awnings, between windows, on walls through the buildings, and on exteriors and interiors of windowsills. Rooftop green designs, as well as expanded tree wells, add more layers of green design. The expanded tree wells include trees, shrubs, tall grasses, and other plants, as well as seating for people to be in the community space or while waiting for public transportation.

Stormwater is the primary source of irrigation, with greywater from the buildings being the secondary source. Water recycling helps lower the cost of the maintenance of the site. Creating as much green, open, and permeable space throughout the neighborhood as possible reduces stormwater damage and cost of repairs. The totality of each area that can use stormwater adds to the total reduction of damage and repairs during a storm event.

The sidewalk is expanded for ease of use for people with strollers, children, people in wheelchairs, people with walkers, and other users to share the space. Colorfully painted sidewalks are wide for ease of walking and are inclusive for use by people with different walking abilities, as well as people with children. This creates an inviting community for a variety of people. Lighting is warm in color and consistent, adding safety throughout the neighborhood for walking, biking, running, or strolling.

Green design adds flowering, edible, and scented plants for physical and emotional health, improving air quality, reducing sound pollution, and creating views of green design from windows. Additionally, the green design creates spaces for birds and other animals that have been displaced due to the development of areas. The effects of the layers of green design for flooding are also beneficial for inclusion in design for a diverse range of people based on their health, abilities, sexuality, gender, age, race, language, income, and culture.

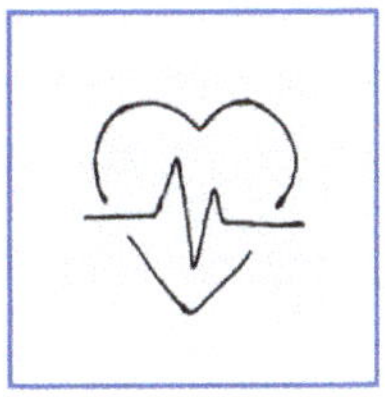

Health – Reducing the effects of stormwater through green design improves people's financial health, emotional health, and physical health. The addition of greening also improves air quality for easier breathing as well as for emotional health. The plants selected produce food, which is shared throughout the community. Street murals use air-purifying paint, which absorbs and neutralizes chemicals, pollutants, and VOCs for improved air quality. Sidewalk seating creates a space for stopping, sitting, or community gathering.

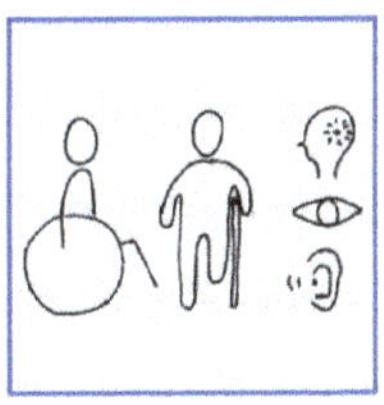

Abilities – A variety of sidewalk seating options are available for people with different mobility skills for healthy walking or while running errands. The street murals are of colors visible to people with different visual abilities. Curb cuts, detectable rumble strips and warning pavers at corners, crosswalk melodies for walking and do not walk, and high-contrast signage improve navigation for people with different visual and walking abilities. With wider sidewalks, there are zones for pedestrians to move at their pace.

Sexuality – Consistent luminescent lighting at pedestrian scale along each block, as well as warm lighting, improves safety for the LGBTQIA+ community. Signage of local businesses welcomes and celebrates people of this community.

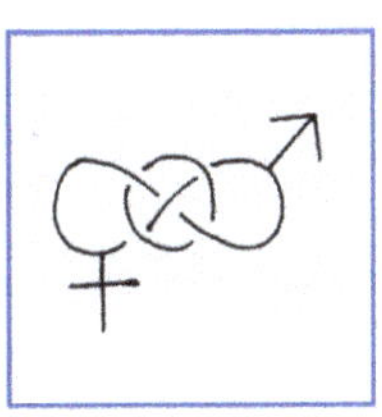

Gender – A variety of sidewalk seating options are available for people with children or who are caregivers for a healthy walking or while running errands. Consistent luminescent lighting at pedestrian scale along each block, as well as warm lighting, improves safety for womxn in the community. Wider sidewalks are spaces for strollers and holding children's hands. Public transportation is routed for womxn and people who run errands and take routes with multiple stops. Designs, signs, and street names celebrate womxn.

Age – A variety of sidewalk designs and seating options are available for young children and mature adults as they are walking or running errands. Curb cuts and safe crosswalks are at every street corner for people to safely cross streets at multiple locations on their route. This is important for people who walk throughout their day while running errands with children and mature people.

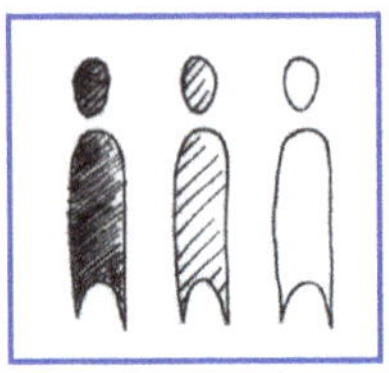

Race – Art, designs, signs, and street names celebrate people of different local, regional, national, and international communities.

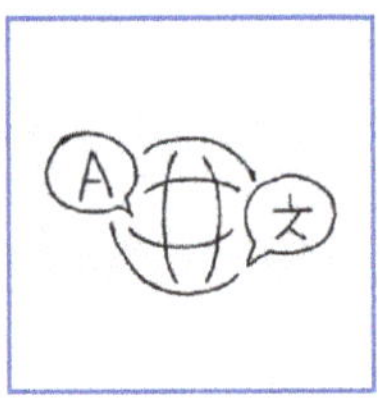

Language – Signs that use multiple local, regional, and international languages, as well as icons and images without words for those learning language, are used throughout the street at pedestrian heights.

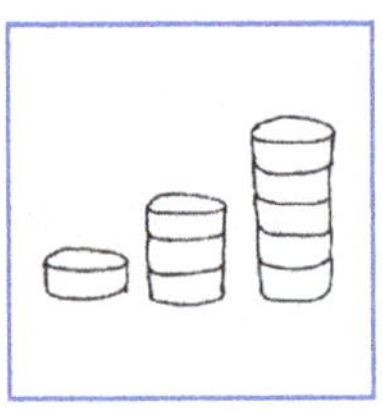

Income – Green designs use rain and stormwater, reducing the effects and costs of storm events. Plants that produce food are included on green rooftops, balconies, green walls, and sidewalk-expanded tree wells. This approach supports people with low income in the community who want or need healthy food. The care and distribution of spaces can also provide jobs to people who want temporary or long-term jobs.

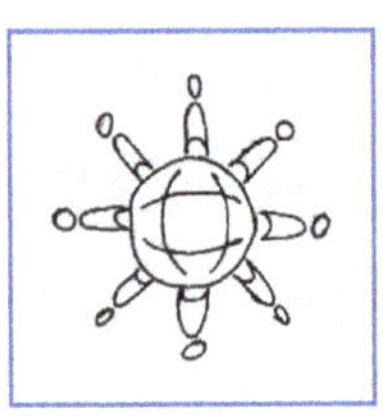

Culture – Food, designs, signs, and street names celebrate people of different local, regional, national, and international communities.

On an urban block scale, greening communities reduces the effects of storm events for the families who live there, the animal population, and overall community health.

Image 3.41 **Green design with both interior and exterior of an urban block with mixed use of residential and commercial spaces.**

The green design reduces rainwater and stormwater using green infrastructure and both vertical and horizontal interventions throughout the community. Green-covered train stations, green-covered bus stops, green roofs, multiple-story green walls, balcony plantings, windowsill planters, and expanded tree wells add living spaces that utilize greywater, rainwater, and stormwater throughout the urban fabric. Green design uses rainwater and stormwater and slows down the water going to the city sewer or combined sewer system.

3.11 Moving Forward with Healthy Designs for Sites That Flood

Reducing stormwater is beneficial, reducing damage and costly repairs as a result of storm events. Severe storms may still cause damage. However, green design can reduce the damage by using the stormwater on site for irrigation. The use of water in green spaces reduces the amount and speed of the stormwater flowing across non-permeable surfaces, which can cause damage or contribute to flooding.

These green site design interventions can reduce stormwater by 10%, 20%, or more, depending on the intensity and length of the rain event. Any reduction of damage is important for structures and cost. The scale of one green design to a house may not seem important to some people. It is, though. When multiple sites add up to a block, neighborhood, and community, the effects are notable. The totality of the green spaces adds up and is effective for reducing damage and the costs of storms.

The effects are damaging to people's lives, people's living situations, and notably are disproportionately affecting people who have low income and who are Black and Hispanic (Mittan, 2020). Green interventions help to improve the inequity of this stormwater damage to people who are marginalized, as well as to a larger community in common, thus addressing and using green design toward a larger picture of inclusion and justice. Adding green design interventions retroactively to site design is a successful approach to inclusion, which has a beneficial ripple effect toward equity and inclusion.

Image 3.42 **Retrofitting spaces, redesigning infrastructure, and repairing protection structures for people to live safely in their community is a long-term approach in which design can support communities of inclusion to be successful and thrive.**

References

Mittan, K. (2020, October 5). Black and Hispanic People More Likely to Live in High-Risk Flood Zones, Study Finds. *University of Arizona News*. Retrieved 20 December 2023, from https://news.arizona.edu/story/black-and-hispanic-people-more-likely-live-high-risk-flood-zones-study-finds

Sasaki Associates. (n.d.). Mississippi River Flood Resiliency Plan. Creating a More Flood Resilient, Accessible, and Beautiful Riverfront for Davenport, Iowa. *Sasaki Associates*. Retrieved 20 December 2023, from https://www.sasaki.com/projects/mississippi-river-flood-resiliency-plan/

Vahedifard, F., Azhar, M., & Brown, D. C. (2023). Overrepresentation of Historically Underserved and Socially Vulnerable Communities Behind Levees in the United States. *Earth's Future*, 11, e2023EF003619. https://doi.org/10.1029/2023EF003619

Wihanesta, R., Sulaeman, D., Pradana, D., & Pribadi, Y. S. (2021). The Reasons for Jakarta's Frequent Flooding and How Nature-Based Solutions (NBS) Can Help Reduce the Risk. *World Resources Institute*. Retrieved 20 December 2023, from https://wri-indonesia.org/en/insights/reasons-jakartas-frequent-flooding-and-how-nature-based-solutions-nbs-can-help-reduce-risk

Chapter 4

Healthy Society and Air Quality

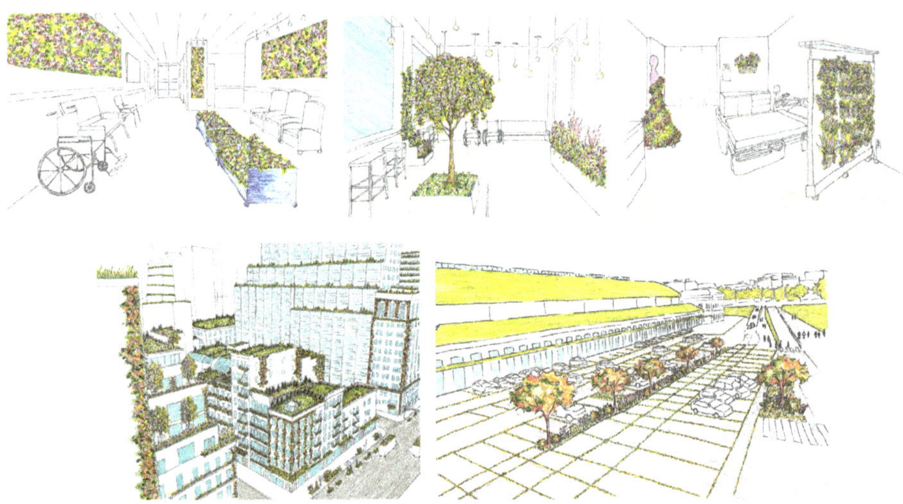

Images 4.1–4.5 **Healthcare facilities, as well as public spaces, are opportunities where green design can aid in improving the physical and emotional health of the people who use the space short or long term.**

 DOI: 10.4324/9781003475293-5

4.1 Introduction

Chapter 4 explores how designs can be improved to address health – physical as well as psychological. Health involves both of these, and green design supports well-being in numerous ways. When designers intentionally create spaces – even if retrofitting is required – for people who are marginalized based on their health, abilities, sexuality, gender, age, race, language, income, or culture, the spaces become inclusive and safer and encourage a greater use of the area.

Healthy air quality includes green designs in spaces to address photosynthesis in which plants change carbon dioxide (CO_2) to oxygen for people and animals to breathe. Further, healthy air also includes removing contaminants that move through the air from industry, vehicles, construction, machinery, materials off-gassing or breaking down, and much more. Contaminants in the air can be removed through a process called phytoremediation, in which plants can uptake pollutants from the air, water, or soil and hold or alter them (without harm to the plants) so that people and animals do not continue to breathe in the pollutants.

Emotional or psychological health is also important for continued growth and a strong society. Emotional health through green design comes through in a variety of layers. Green designs in healing centers such as hospitals or therapy sites for people to see living plants while they are recovering improves their emotional and even their physical well-being (Odom, 2016; Raanaas et al., 2011; Wells and Evans, 2003; Wolf and Flora, 2010; World Health Organization, 2016).

Further, green design can improve community health with designs that bridge and reconnect people to spaces that have been separated due to zoning, poor road design, and policies that reinforced segregation (Allen, 2023; Evans, 2021). As the ability to address and change policies is slow, and roads and highways that segregate communities may not be torn down anytime soon (if at all), the use of green design to create healthy air in these communities, as well as creating greenbelts to connect the communities to their surroundings, is a healthy green design solution. Additionally, this approach acknowledges some of the issues while exploring a solution toward a greater goal of health.

Studies in this chapter explore how green design can promote health for people who experience post-traumatic stress disorder (PTSD), as well as plants that promote improved air quality for health and healthy food (which is explored more in Chapter 2).

4.2 Healthcare Facility Corridor Green Design for Patients, Families, Caregivers, and Staff

The green design for a healthcare facility is created for an ever-changing space. In interviews with numerous users of healthcare facilities, including short-term and long-term patients, staff, and healthcare professionals, the need for interior green design was expressed throughout the dialogues.

The design illustrated in Image 4.7 creates a walking route for patients in recovery, clearly labeled, with options for seating, fresh scent, and the ability to experience and see healthy changing design – such as plants growing. This is notable for patients who are in long-term recovery and stay for days or weeks. Staff and professionals also require time to sit and emotionally recover. The ability to see, smell, and touch a continually growing green design for workers who are at the facility on a daily basis is a way of experiencing growth, health, and time moving forward with an aesthetic beauty.

This design includes seating at different heights, shapes, abilities, and comfort levels. Inclusive seating includes those for children and adults who need electronic support to sit down or stand up. There is space for people who are in wheelchairs to easily move. Comfortable inclusive seating encourages people to spend time experiencing the light fresh scent of the green design – the plant lights of the spaces.

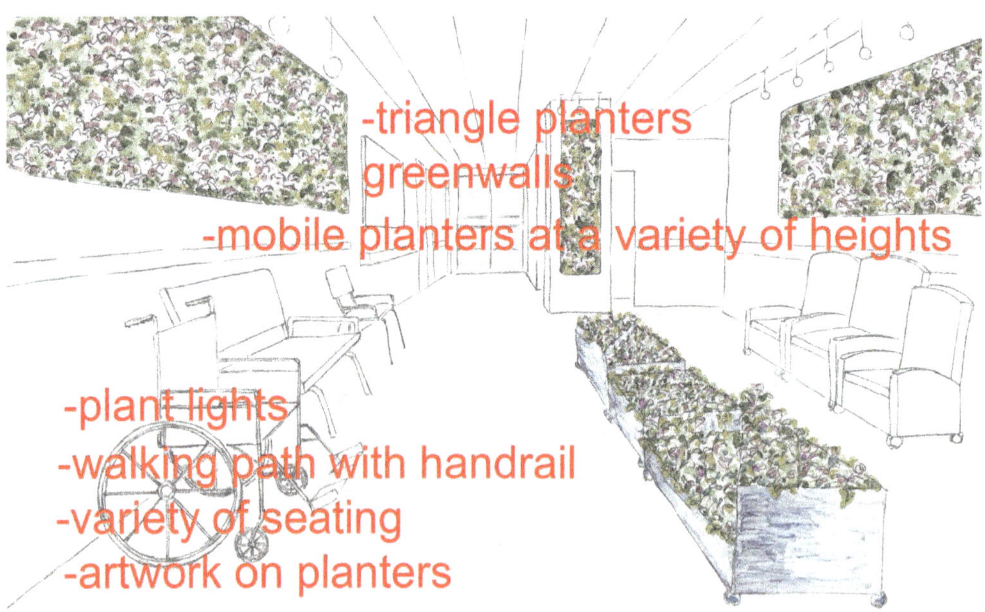

Image 4.6 Including a space for people in recovery, as well as those who work at healthcare facilities, supports an inclusive space of green design for health for people no matter their income.

Edges of the seating, planters, and full design are soft for safety. Mobile planters are three-sided, allowing them to be arranged in a variety of spaces more easily. The artwork on the planters can be created by patients, artists, or the community. The designs on the planters can also change. Mobile planters allow the space to more easily change when needed, moved to patients' rooms, or moved for needed lights, watering, or other care needs. With planters of different heights, people of different heights are able to experience the greening that is meant for them. If wanted, staff, professionals, and patients can be trained to care for some of the living space – which is soothing to some people.

Living green vertical artwork includes flowering and leafy plants. Lighting and irrigation are designed into the space specifically for the green walls for long-term growth. Seeing the space change on a daily, weekly, or monthly basis as the plants grow gives people a minute to pause and see life grow in front of them. Experiencing healthy growth in a green design is not only psychologically healthy in healthcare settings – it is also a beautiful experience that can be shared by strangers and friends alike.

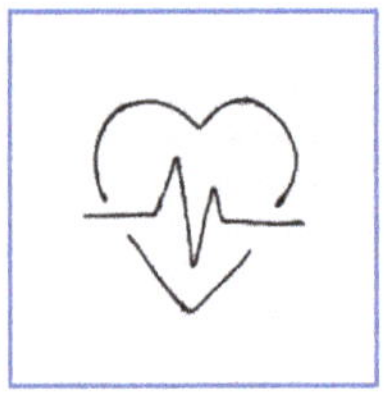

Health – Create colorfully marked paths to know directions for walking, the distance of the path, and views along it. The living green design improves people's emotional and physical health. There are designed zones or areas for enjoying sitting with views of interior and exterior green design. There are also paths for people to walk and move machinery. The green design improves air quality for easier breathing as well as for emotional health.

Abilities – The planters and living designs are different heights, allowing people with different physical abilities to interact with the plants. With the planters being mobile, the space can easily be changed to account for people with different physical needs. The seats have a variety of seating styles, heights, and options for mixed users.

Sexuality – The artwork on the planters celebrates the LGBTQIA+ community. Further, in extending the applications of the SDOH, in which design of spaces is an important determinant of health, this design also applies healthy greening and other components for health equity for people who are and have been marginalized through a range of areas such as health, abilities, sexuality, gender, age, race, language, income, and culture.

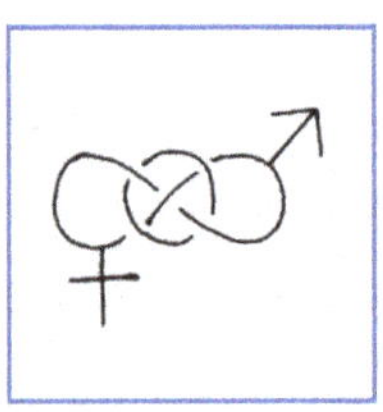

Gender – The artwork on the planters celebrates multiple genders. In extending the applications of the SDOH, in which design of spaces is an important determinant of health, this design also applies healthy greening and other components for health equity for people who are and have been marginalized through a range of areas such as health, abilities, sexuality, gender, age, race, language, income, and culture.

Age – The planters are of different heights, allowing people of different heights to enjoy the scent, sight, and textures of the green space.

Race – In extending the applications of the SDOH, in which design of spaces is an important determinant of health, this design also applies healthy greening and other components for health equity for people who are and have been marginalized through a range of areas such as health, abilities, sexuality, gender, age, race, language, income, and culture.

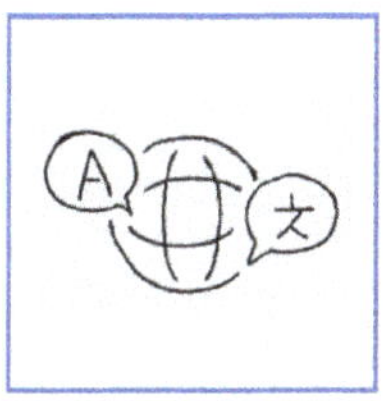

Language – Create paths and spaces where icons and colors are used to clearly navigate to find locations within the facility. High-contrast signs with multiple languages and icons in the space at pedestrian scale.

Income – Green design improves emotional and physical health and therefore can help patients, caregivers, and staff in healthcare facilities, no matter the income level of the patients and public healthcare facility. Green design should not be considered a luxury for higher-income facilities but rather a source to improve the health of people, no matter their income.

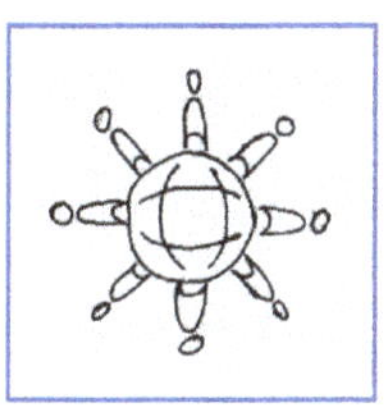

Culture – The artwork on the planters celebrates multiple cultural identities. Further, in extending the applications of the SDOH, in which design of spaces is an important determinant of health, this design also applies healthy greening and other components for health equity for people who are and have been marginalized through a range of areas such as health, abilities, sexuality, gender, age, race, language, income, and culture.

The shared green space in the hospital or healthcare facility is an opportunity to create a design to enhance and enjoy the company of others for people in both short-term and long-term stays. Further, the experience of a living space improves the emotional and physical healing of the patients and caregivers.

Image 4.7 **Healthcare facility shared green space for patients, families, caregivers, and staff.**

This healthcare facility green design is in a shared space for patients, their families, caregivers, and staff. It has mobile green planters at varying heights, as well as vertical greening. The changing and growing colorful green design in the space, as well as green views from the windows, includes comfortable seating for a variety of people. Plant lights are run on a timer for the green design, as well as comfortable lighting for people's eyes.

Images 4.8–4.9 The Green Road Project at Walter Reed National Military Medical Center is located in Bethesda, Maryland, United States. Green design is used to promote healing of wounded warriors with the invisible wounds of war.

4.3 The Healing Effects of Nature for PTSD, Bethesda, Maryland, United States

Scientists are studying the healing effects of nature mathematically through the Green Road Project, in initiative of the Institute for Integrative Health (Odom, 2016). The study focuses on the effects of natural living green settings on people with traumatic brain injuries, as well as with PTSD and other psychological conditions.

The research will take a few years, said Fred Foote, a retired U.S. Navy neurologist who works with the Institute for Integrative Health. However, the researchers quickly found that when veterans are in a green woodland, they quickly turn happy and perk up and have a great time with PTSD symptoms set aside for the time while they are in the green woodland.

Images 4.10–4.11 The Green Road Project at Walter Reed National Military Medical Center is for people in the military with PTSD and other conditions and wounds. The forested space has paths, private spaces, and gathering spaces for small groups of people to sit within the changing and growing care facility.

Image 4.12 **The Green Road Project at Walter Reed National Military Medical Center has a space that changes colors throughout the year as the temperature and sunlight change. Healthy green visible change aids emotional healing.**

The 1.7-acre woodland garden is located at Walter Reed National Military Medical Center in Bethesda, Maryland. Multiple studies have found that in a natural space, people can reduce stress, and healing is sped up, especially those with PTSD. This study is different in that it measures isolated biomarkers with quantitative outcomes instead of self-reported research of study participants.

Image 4.13 **Creating pathways through green design for users with different abilities is an inclusive approach in design for diverse users to share green spaces.**

The application of green design for PTSD has further applications for others who experience this condition. In research by Webb and colleagues (2024), green spaces in neighborhoods were found to be positively affecting people with PTSD. Approximately 2,600 people were recruited from emergency visits across the United States within 72 hours of the trauma. The individuals self-reported that with green space and other forms of support, they had low to no symptoms over time.

The National Recreation and Parks Association (n.d.) writes that people living more than 1 kilometer (0.62 miles) from a green space have 50% higher odds of experiencing stress than those living less than 300 meters (0.19 miles) from a green space.

4.4 Healthcare Therapy Room Green Design

The physical therapy room in a healthcare facility illustrated in Image 4.15 is a green design of a shared space for patients and caregivers. Creating a planted design in a room in a healthcare facility for therapy includes an interior tree, a trellis for vertical greening, planters with plants that may be scented, seating, and plant lights. The room is a garden using living design as its artwork.

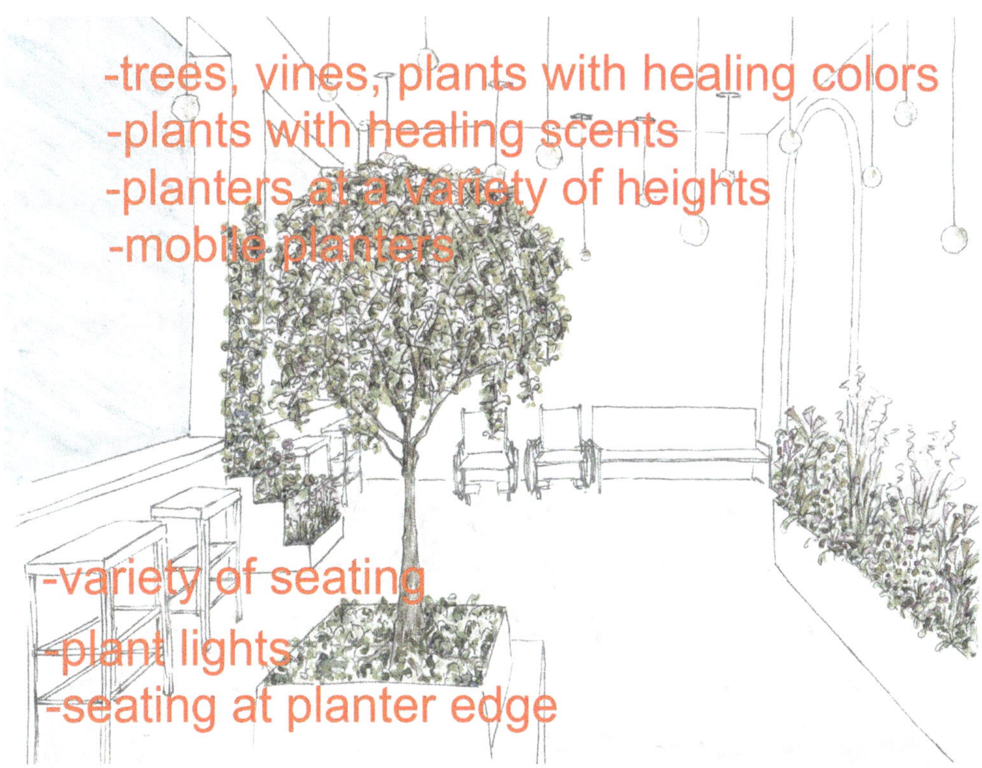

Image 4.14 Including a planted room for people in recovery, as well as those who work at healthcare facilities, supports an inclusive space for people of mixed income levels.

Plant lights are run on a timer for the green design, as well as comfortable lighting for people's eyes. The plant lights, which are at varying heights throughout the space, can be set on a timer to turn on with a sunrise and dim with sunset. Room lights allow users to enjoy the space throughout all suitable hours.

Having an indoor tree that grows in the therapy room can show visible change over time, reminding people that they themselves also grown and change. The changing and growing colorful green design in the space, as well as the green views from the windows can be enjoyed from comfortable seating for a variety of people. The tree planter has a structure for seating. Further mobile seating is included for children, adults, and people with different mobility abilities.

A trellis has plants from an adjacent planter growing through the structure. The layers of plants can be leafy green, have colorful leaves, or have flowers. The irrigation may be silent or have a relaxing, soothing sound with the water moving through the green sculptures. Care for the green room includes professionals, volunteers, and patients. Care for the garden can improve physical health with people who need to practice motor skills, as well as psychological health for people to experience green spaces.

This design addresses health as a basic right for people, no matter if their income is low or high. Physical and psychological health in care facilities through green designs is open for all who need the healing space. This space is for use by patients, staff, and professionals.

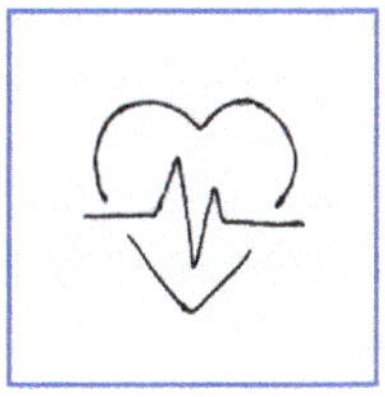

Health – Using a green design in a physical therapy room creates a changing and growing space. As people physically heal, they can further have emotional growth. Experiencing plants as the garden changes and grows reminds people in the healing process that they also heal and grow. In addition to using green design to improve air quality for easier breathing, emotional health is also employed.

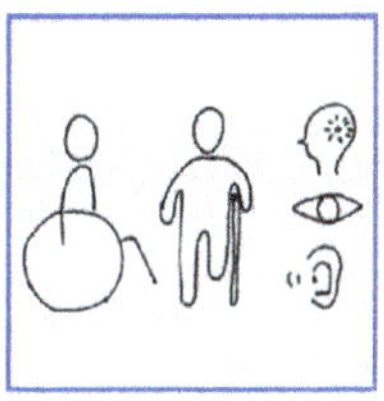

Abilities – The planters are of different heights, allowing people with different physical abilities to interact with the plants. Scented plants allow users with different visual abilities to enjoy the room. The green design has mobile vertical plantings, allowing space for people with different walking abilities to safely navigate through the space or relax in the room.

Sexuality – The artwork on the planters celebrates the LGBTQIA+ community. Further, in extending the applications of the SDOH, in which design of spaces is an important determinant of health, this design also applies healthy greening and other components for health equity for people who are and have been marginalized through a range of areas such as health, abilities, sexuality, gender, age, race, language, income, and culture.

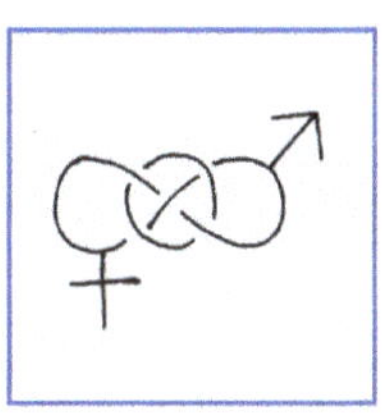

Gender – In extending the applications of the SDOH, in which design of spaces is an important determinant of health, this design also applies healthy greening and other components for health equity for people who are and have been marginalized through a range of areas such as health, abilities, sexuality, gender, age, race, language, income, and culture.

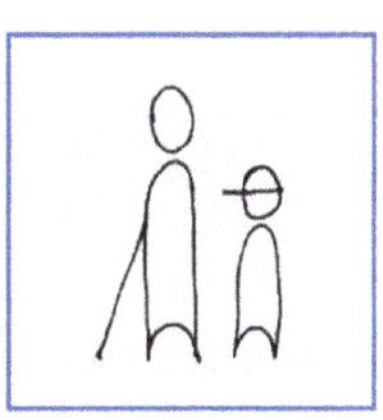

Age – The green design and planters are of different heights and different styles, allowing people of a variety of ages, heights, and wants to enjoy the scent, sight, and textures of the green space.

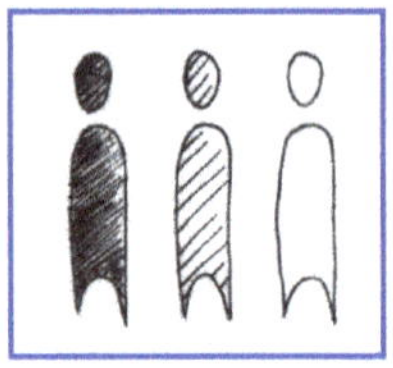

Race – In extending the applications of the SDOH, in which design of spaces is an important determinant of health, this design also applies healthy greening and other components for health equity for people who are and have been marginalized through a range of areas such as health, abilities, sexuality, gender, age, race, language, income, and culture.

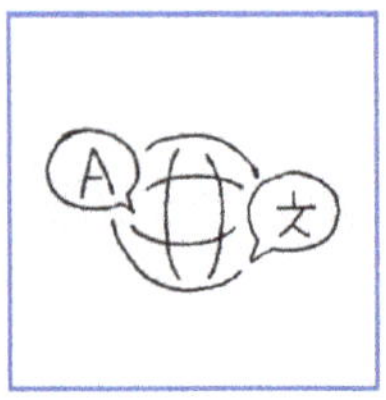

Language – High-contrast signs with multiple languages and icons in the space at pedestrian scale.

Income – Green design improves emotional and physical health and therefore can help patients, caregivers, and staff in healthcare facilities, no matter the income level of the patients and public healthcare facility. Green design should not be considered a luxury for higher-income facilities but rather a source to improve the health of people, no matter their income.

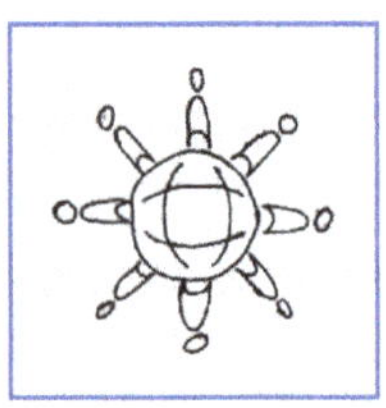

Culture – The artwork on the planters celebrates multiple cultural identities. Further, in extending the applications of the SDOH, in which design of spaces is an important determinant of health, this book also applies healthy greening and other components for health equity for people who are and have been marginalized through a range of areas such as health, abilities, sexuality, gender, age, race, language, income, and culture.

The shared green physical therapy space in the healthcare facility is an opportunity to create a design to enhance the company of others for people in both short-term and long-term stays. Further, the experience of a living space improves the emotional and physical healing of the patients and caregivers.

Image 4.15 **Physical therapy room in a healthcare facility.**

This physical therapy room in a healthcare facility green design is in a shared space for patients and caregivers. Having an indoor tree that grows in the therapy room can show visible change over time, reminding people that they themselves also grow and change. The changing and growing colorful green design in the space, as well as green views from the windows, has comfortable seating for a variety of people. Plant lights are run on a timer for the green design, as well as comfortable lighting for people's eyes.

Image 4.16–4.17 **Dense urban areas without green spaces have poor air quality.**

4.5 A Greener Future for Everyone: Green Space Inequity in San Francisco

In 2017, San Francisco became the first city in the United States to intentionally develop a design that ensures all its residents are within a ten-minute walk of some type of park, playground, or green space (Ajzurcher, 2024). This is an approach that has great intentions and outcomes for the residents.

This, however, does not mean that all the spaces are equitable. For example, in San Francisco people in low-income and Black, Brown, and underserved populations, there is less green space than there is for those of Caucasian populations.

- Neighborhoods of predominantly Black, Brown, and underserved communities have 35% fewer parks than the average neighborhood and 56% less park space than the average neighborhood that is predominantly Caucasian.

Image 4.18–4.19 **The addition of green design to urban areas in large swaths of spaces, as well as multiple smaller spaces, weaves a greenbelt of health throughout a community to improve not only air quality but also healthy green gathering spaces.**

Image 4.20 Small spaces, green roofs, living vertical designs, pocket parks, and large green spaces weave a mix of public green spaces for communities to share. These are important for the physical health of people through improved air quality, as well as the emotional health of being grounded in earthy spaces while enjoying the density of urban design.

- People in low-income neighborhoods have 38% less park space than the average neighborhood and 58% less than that found in high-income neighborhoods.

There are similar patterns when looking at urban forest and tree coverage in San Francisco. Neighborhoods with the lowest percentage of tree cover have high rates of poverty and racial minority population. Wealthy neighborhoods have a high percentage of tree cover.

Image 4.21 The addition of green design woven throughout an urban fabric improves the quality of life for the residents, workers, and visitors.

Developing equitable green space and programs can start from multiple directions – from the top down, and from the grassroots up. Both approaches have been taking place to address the green space inequity in San Francisco: from the top down – in 2020, 80% of the capital budget for parks went to projects in underserved communities – and from grassroots – a shared schoolyard program works with local schools to make playgrounds available after school hours for families and community members.

Working toward green space equity requires parties from different directions and scales to work together find small-scale and large-scale solutions for communities to have access to parks, playgrounds, and green spaces. With people speaking up, as well as making sure that people listen, and then follow through with solutions, equitable green space access is achievable. With equitable green space the physical and emotional health of the community is improved.

4.6 Patient Room Green Design

Creating a green design in a patient's room, as illustrated in Image 4.23, includes using self-contained mobile living green sculptures and green walls, as well as wall planters. The wall planters – which include fresh-cut flowers or growing plants to provide changing color, changing scenes, and fresh scents – are attached to the wall.

Image 4.22 **Self-contained mobile living artwork that can be moved to different rooms as patient conditions change is one approach to experiencing green design in rooms without windows, as well as with patients who are unable to move from their room.**

Green mobile designs have a variety of styles and sizes of green sculptures created for different patients and rooms. Each living piece has its own set of plant lights and irrigation system to create a self-contained piece that can be moved easily. The irrigation may be almost silent or have a relaxing, soothing sound with the water moving through the green sculptures.

The mobile green wall has plants that can be leafy green, have colorful leaves, or have flowers. The planter pockets can have changing plants with lovely scents that can be changed easily as needed. The mobile green wall can create semi-privacy for patients, friends, and family.

Mobile garden walls, green sculptures, green fashions, living artworks, and other mobile pieces are unique and can be moved to be cared for by professionals or by volunteers who enjoy gardening. This can improve physical health with people who need to practice motor skills, as well as psychological health for people to experience green spaces.

Exposure to green designs – both through a window or in the same room – improves emotional well-being, improves recovery rates in health, and creates spaces and designs that change over time with new growth of leaves and flowers. These changes improve the health of people. Creating mobile designs is an approach to bring greening to spaces that do not have windows and need to change based on patients and caretaker needs (American Chemical Society, 2009; Raanaas et al., 2011; Wells and Evans, 2003; Wolf and Flora, 2010; World Health Organization, 2016).

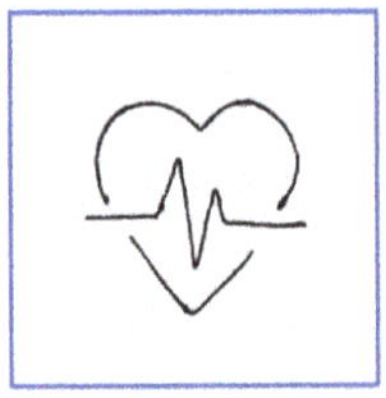

Health – Using a green design in a patient's room creates a changing and growing space; as people physically heal, they can further have emotional growth. Experiencing plants as they change and grow internally reminds people in a healing process that they also heal and grow. In addition to using green design to improve air quality for easier breathing, emotional health is also improved with green design in this space.

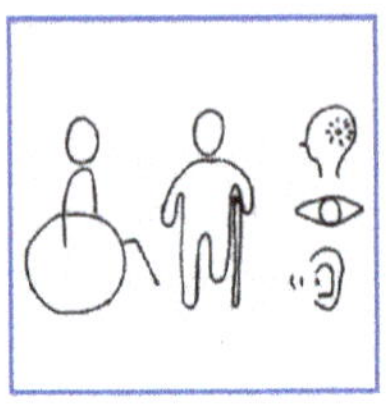

Abilities – The planters are of different heights, allowing people with different physical abilities to interact with the plants. Scented plants allow users with different visual abilities to enjoy the room. The green design has notable mobile designs, allowing people with different mobility and vision to navigate through the room well.

Sexuality – In extending the applications of the SDOH, in which design of spaces is an important determinant of health, this design also applies healthy greening and other components for health equity for people who are and have been marginalized through a range of areas such as health, abilities, sexuality, gender, age, race, language, income, and culture.

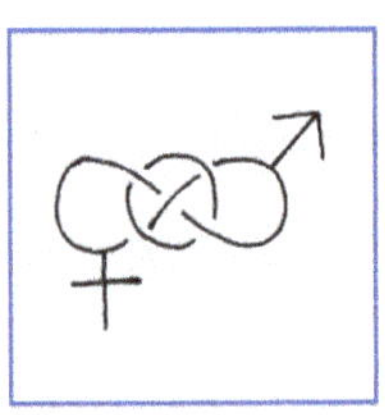

Gender – In extending the applications of the SDOH, in which design of spaces is an important determinant of health, this design also applies healthy greening and other components for health equity for people who are and have been marginalized through a range of areas such as health, abilities, sexuality, gender, age, race, language, income, and culture.

Age – The mobile living sculptures' green design and planters are of different heights and different styles, allowing people of a variety of ages, heights, and wants to enjoy the scent, sight, and textures of the green space.

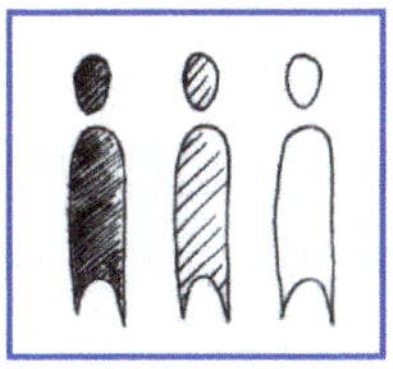

Race – In extending the applications of the SDOH, in which design of spaces is an important determinant of health, this design also applies healthy greening and other components for health equity for people who are and have been marginalized through a range of areas such as health, abilities, sexuality, gender, age, race, language, income, and culture.

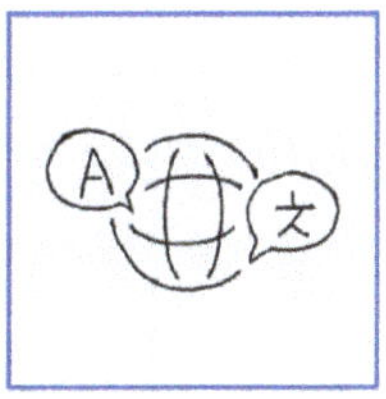

Language – High-contrast signs with multiple languages and icons in the space at pedestrian scale.

Income – Green design improves emotional and physical health and therefore can help patients, caregivers, and staff in healthcare facilities, no matter the income level of the patients and public healthcare facility. Green design should not be considered a luxury for higher-income facilities but rather a source to improve the health of people, no matter their income.

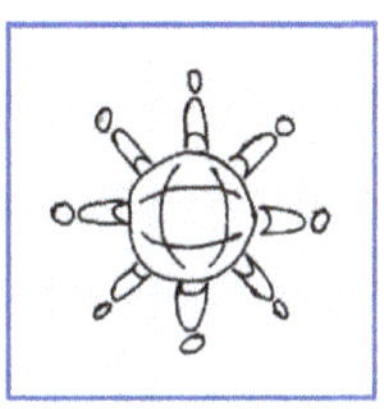

Culture – In extending the applications of the SDOH, in which design of spaces is an important determinant of health, this design also applies healthy greening and other components for health equity for people who are and have been marginalized through a range of areas such as health, abilities, sexuality, gender, age, race, language, income, and culture.

A patient's room in a healthcare facility is an opportunity for creating a design for people in both short- and long-term stays. The practice of viewing and experiencing green design improves the emotional and physical healing of the patients and caregivers.

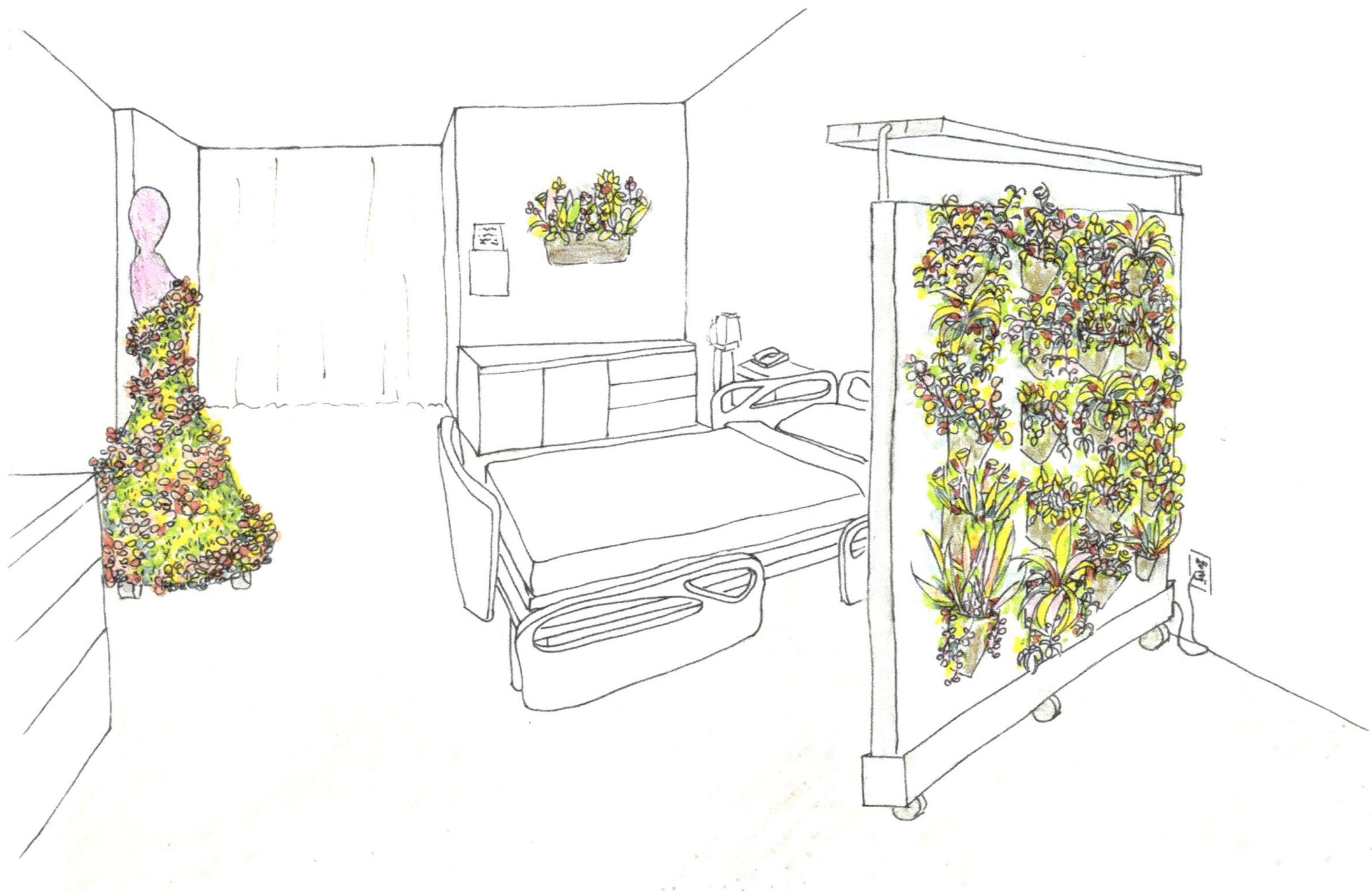

Image 4.23 Patient room in a healthcare facility.

The patient's room has movable green sculptures; living artwork walls are self-contained with lights and an irrigation system. The sound from the moving water pumping through the green design can be soothing. The wall planter with freshly cut flowers or growing plants provides changing color, changing scenes, and fresh scents.

Images 4.24–4.25 **Dense urban areas without green spaces have poor air quality, which could be improved, in part, with the use of green design retrofitted into the community.**

4.7 Gentrification Leads to Worsened Air Quality, Detroit, Michigan, United States

Research from Henry Ford Health and Michigan State University looks at air quality in Detroit, Michigan after gentrification (Swanson, 2023). After building three sports venues, new restaurants opened, as did bars, hotels, and large residential complexes – and the air quality worsened. Increased construction and increased traffic volumes produced increased exhaust. Further, poor air quality is more prominent during rush hour and night-time, as well as during the winter and summer – when air is more stagnant.

Dr. Ikenna Okereke, M.D. and Dr. Qiong Zhang, Ph.D. are working to add hundreds of air quality sensors to monitor air quality in Detroit and develop a mobile application with which people can check the quality of their air in real

Image 4.26–4.27 **The addition of green design to urban areas in large swaths of spaces, as well as multiple smaller spaces, weaves a greenbelt of health throughout a community to improve not only air quality but also healthy green gathering spaces.**

Image 4.28 **Density of diverse populations and access to shared public spaces and public comforts are why urbanites enjoy dense spaces. Healthy green space for people who choose to live here is one approach, as health is a basic right for people.**

time. Poor air quality is damaging to people's health with an increased risk for lung cancer, heart disease, lung diseases such as pulmonary fibrosis and asthma, preterm birth, and decreased immune health and emotional health.

Emotional health is affected because poor air quality is likely linked to lack of public green spaces, and without access to green spaces, children and adults are more likely to develop emotional issues.

Image 4.29 **Small spaces, green roofs, living vertical designs, pocket parks, and large green spaces weave a mix of public green spaces for communities to share. These are important for the physical health of people through improved air quality, as well as the emotional health of being grounded in earthy spaces while enjoying the density of urban design.**

Without the addition of green spaces as an integral required piece of developing an area, the short-term and long-term effects of gentrification of areas will continue to include poor air quality.

As areas become gentrified, green spaces and green infrastructure are not expanded. This affects the air quality, which affects people's health both physically and emotionally. Another important effect of gentrification is that people are displaced from their homes and communities. Gentrification disproportionately affects minorities (Feder, 2020).

In the city of Philadelphia,

> Gentrification is reconfiguring the urban landscape by shrinking residential options within cities for disadvantaged residents and expanding them for more advantaged residents. Our findings suggest that, for the Black community, there are additional constraints when they move, leading them to move to a shrinking set of affordable yet disadvantaged neighborhoods within the city.
>
> (Ding and Hwang, 2020)

Investing significantly in non-gentrifying neighborhoods would:

> ensure that disadvantaged movers are not limited to neighborhoods with high levels of disadvantage, high crime and low-quality schools.
>
> (Ding and Hwang, 2020)

4.8 Urban Community Green Design

Green design weaves throughout this urban neighborhood in Image 4.31. Green interventions – both small and large, vertical, and horizontal, and on private and public sites – weave through this site design. The total effect is seen and experienced by pedestrians, drivers, and people who use public transportation and from window views by residents, workers, and visitors and improves the physical air quality and emotional health of the people of the community as they smell fresh air and see change through the growing design.

The green design is weaved on rooftops, balconies, green walls, algae biopanels, windowsills, awnings, expanded street tree wells, sidewalk planters, protected bike lanes with green planters, and buses with planted roofs. Greywater and stormwater irrigate the green design throughout the neighborhood.

Green design in urban areas uses stormwater for irrigation for plant growth. Using stormwater in urban spaces through greening reduces the amount of water that arrives at storm sewers or storm drains, reducing the effects of flooding. Further, the green design filters stormwater for improved quality before it gets to the water table. Greywater is also available for irrigation from the adjacent commercial and residential buildings.

The totality of the design is created with both small and large green interventions in both public and private spaces. The benefits of the design can be experienced by users of the site – both residents and workers. The physical benefits of the design include improved air quality including reducing particulate matter PM 2.5 in the air. Plants reduce particulate matter, filtering contaminants in the air, and produce oxygen for people to breathe.

Inclusivity in this design includes pedestrian-scale warm lighting throughout the space, which adds safety, particularly for womxn and the LGBTQIA+ community. Seating for a variety of users, as well as art celebrating a variety of cultures and races, are celebrated in this design.

The equity and inclusion in this green design of urban spaces addresses the ideal that all people have a right to health, including the health provided by green spaces. The green design provided here retrofits an existing neighborhood with living spaces without tearing down existing architecture. Creating green design in neighborhoods, such as the one illustrated here, requires seeing the space with fresh eyes for weaving unique living design in smaller interventions, adding to the total space. In neighborhoods where green spaces have been lacking due to inequitable and exclusive design practices, creative green design opportunities must be explored at multiple layers to add to the totality of the space for improving people's physical and emotional health.

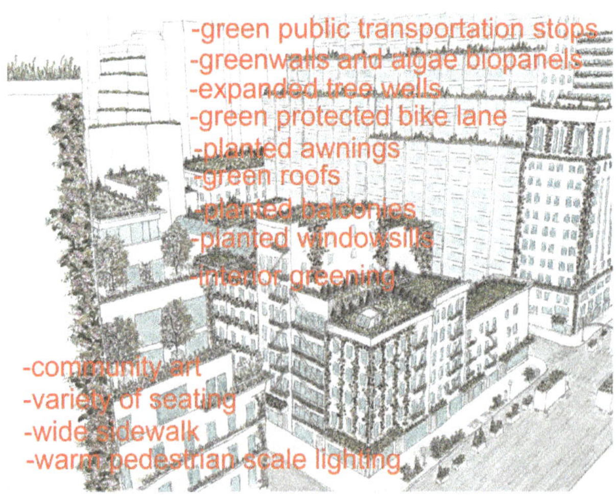

Image 4.30 Urban neighborhood with green design. The layers of green interventions weave throughout the space in both small and medium scale, improving health and equity.

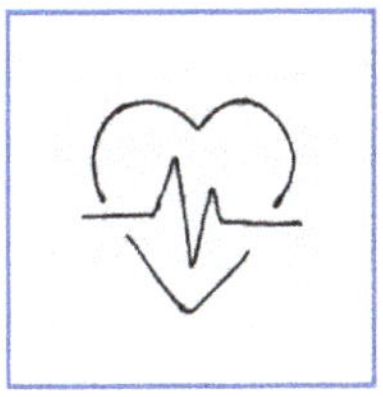

Health – Views from windows benefit users who are not easily able to experience other ways. Further, in extending the applications of the SDOH, in which design of spaces is an important determinant of health, this design also applies healthy greening and other components for health equity for people who are and have been marginalized through a range of areas such as health, abilities, sexuality, gender, age, race, language, income, and culture.

Abilities – Views from windows benefit users who are not easily able to experience other ways. Further, in extending the applications of the SDOH in which design of spaces is an important determinant of health, this design also applies healthy greening and other components for health equity for people who are and have been marginalized through a range of areas such as health, abilities, sexuality, gender, age, race, language, income, and culture.

Sexuality – Luminescent lights and warm lights improve the safety of the space for womxn, the LGBTQIA+ community, and other users. Designs, signs, and street names celebrate the LGBTQIA+ community.

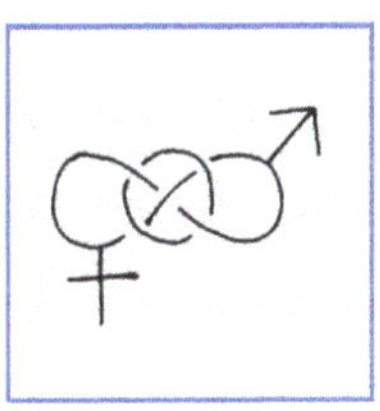

Gender – Luminescent lights and warm lights improve the safety of the space for womxn, the LGBTQIA+ community, and other users. Designs, signs, and street names celebrate the LGBTQIA+ community.

Age – In extending the applications of the SDOH, in which design of spaces is an important determinant of health, this design also applies healthy greening and other components for health equity for people who are and have been marginalized through a range of areas such as health, abilities, sexuality, gender, age, race, language, income, and culture.

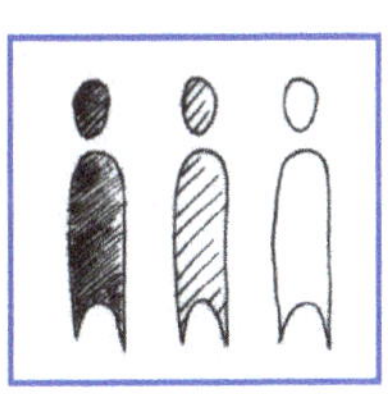

Race – In extending the applications of the SDOH, in which design of spaces is an important determinant of health, this design also applies healthy greening and other components for health equity for people who are and have been marginalized through a range of areas such as health, abilities, sexuality, gender, age, race, language, income, and culture.

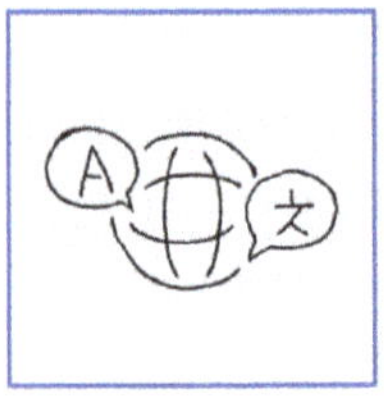

Language – Art and signs in a variety of languages and universally understood icons are designed throughout the site.

Income – Plants that produce food are put in public spaces for people who want or need it. Views from windows benefit users. Further, in extending the applications of the SDOH in which design of spaces is an important determinant of health, this design also applies healthy greening and other components for health equity for people who are and have been marginalized through a range of areas such as health, abilities, sexuality, gender, age, race, language, income, and culture.

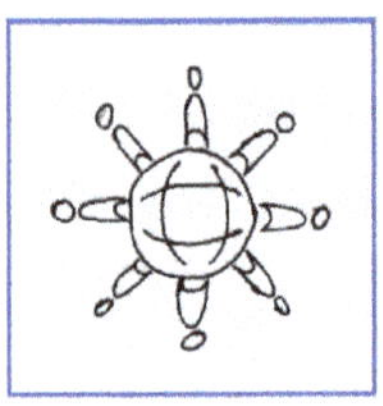

Culture – Art, designs, forms, materials, and uses of the green spaces can be culturally focused. Further, in extending the applications of the SDOH in which design of spaces is an important determinant of health, this design also applies healthy greening and other components for health equity for people who are and have been marginalized through a range of areas such as health, abilities, sexuality, gender, age, race, language, income, and culture.

Integrating green design throughout an urban space creates connecting communities with improved physical and emotional health, improved air quality for the health of people and animals, and green spaces for animal habitats.

Image 4.31 Green design in an urban neighborhood.

Green interventions – small and large, vertical and horizontal, and on private and public sites – weave through this site design. The total effect is seen by pedestrians, drivers, people who use public transportation; from window views for residents, workers, and visitors; and improved physical air quality and emotional health of the people of the community.

Image 4.32 **Photos of the existing highway system help to explain how the road structures were used to harm and displace people.**

4.9 How Interstate Highways Gutted Communities – and Reinforced Segregation

The racism involved in the Highways Act of 1956 is discussed by Evans (2023). Planners claimed that their goal was to eliminate unsafe roads and inefficient routes, and to create safe cross-country travel by creating 41,000 miles of roads. However, their other goal – eliminating Black communities – was evident in their words and actions.

Image 4.33 **Highways have been designed to separate and remove people from communities with a specific target of eliminating Black communities (Evans, 2023).**

Image 4.34 **Retrofitting highways to address connections between the separated communities is an inclusive approach to design.**

With 475,000 households and more than a million people displaced nationwide, the majority of the people targeted in this statistic were Black and low-income communities. The Federal-Aid Highway Act of 1956 caused a lack of pedestrian paths through communities; communities being cut off from other areas; worsened air quality; significantly declined property values; loss of churches, green spaces, homes, and jobs; and physical walled barriers of roads and walls of concrete acting to enforce segregation – all done through racist zoning policies and de-investing in these neighborhoods.

Robert Moses – one of the powerful people involved with highways, roads, bridges, parks, and housing nationally in the United States – led the practice of "the best way to eradicate the supposed slums where Black people lived was to build highways through them" (Evans, 2023). In 1959, Moses said: "Our categorical imperative is action to clear the slums."

The Federal-Aid Highway Act of 1956 enforced that the federal administration would pay for 90% of the cost of the state's new roads – but only if the state agreed to run roads through the major city to the suburbs and remove undesirable populations. Highway engineers actively participated in this practice. Black and low-income neighborhoods were predominantly targeted, as they were considered undesirable by politicians and the impetus for white flight.

As civil rights were gaining motion in the 1950s, the Highway Act was a way to disseminate the Black communities and create physical road barriers to recreate and reinforce the then-recently removed segregation practices. Practices included either cutting directly through Black communities with the highways or encircling a Black community to confine residents with a maze

Image 4.35 **Adding green spaces throughout areas with roadways, including nearby and woven under highways, is an approach for healthy shared community spaces.**

of barriers, roads, and walls. This knowingly and intentionally created an area of poor air quality due to the construction and roads, lack of funding for the area, lack of public transportation, and lack of open green spaces for residential health.

Affected were major cities including New York, Miami, Chicago, Minneapolis, Pittsburgh, Oakland, Nashville, Baltimore, Atlanta, and more. For example, Interstate 81 in Syracuse was built to destroy the 15th Ward, a Black community that city planners believed was too close to Syracuse University and the growing downtown area. In Miami, Black residents that who were displaced by the highways were blocked from moving into white neighborhoods. The displaced Black residents were forced into overcrowded areas of the city struggling from urban decay, poverty, and poor living conditions.

4.10 Urban Parking Plaza Green Design

Urban parking plazas are opportunities to weave green infrastructure through an urban fabric. Green design improves physical and emotional health. The totality of the green interventions in the design in Image 4.37 encourages community gathering through weaving greening through the neighboring communities.

The totality of the design is created with both small and large green interventions in public and private spaces. The benefits of the design can be experienced by users of the site – both residents and workers. The physical benefits of the design include improved air quality, including reducing particulate matter PM 2.5 in the air. Plants reduce particulate matter, filtering contaminants in the air, and produce oxygen for people to breathe.

This design includes permeable pavers and bioswales in the parking lot, as well as green roofs throughout the neighborhood. Green walls and algae biopanels are designed on vertical surfaces from short walls through multiple-story buildings. Street trees and expanded tree wells are further planted within the neighborhood to improve air quality for physical health, as well as emotional health due to growing green designs.

Inclusivity in this design includes pedestrian-scale warm lighting throughout the space for added safety, particularly for womxn and the LGBTQIA+ community. Seating for a variety of users, as well as art celebrating a variety of cultures and races, is celebrated in this design.

Green design in urban areas uses stormwater for irrigation for plant growth. Using stormwater in urban spaces through greening reduces the amount of water that arrives at storm sewers or storm drains, reducing the effects of flooding. Further, the green design filters stormwater for improved quality before it gets to the water table. Greywater is also available for irrigation from the adjacent commercial and residential buildings.

Parking structures have been a blight on design, sometimes by blocking views, and separating neighborhoods of people with low income, as well as for racial and cultural divisions. Through an integrated green design, urban sounds are reduced and health is improved, while opening up spaces that are inclusive and safe for a larger diverse population.

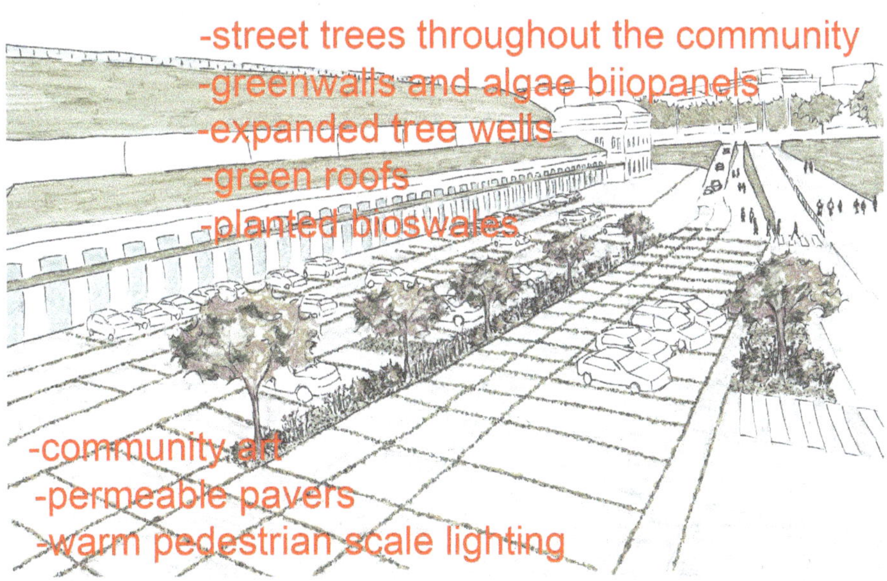

Image 4.36 **Green parking plazas with layers of greening. Creating a safe green inclusive plaza includes looking at small-scale and large-scale design changes.**

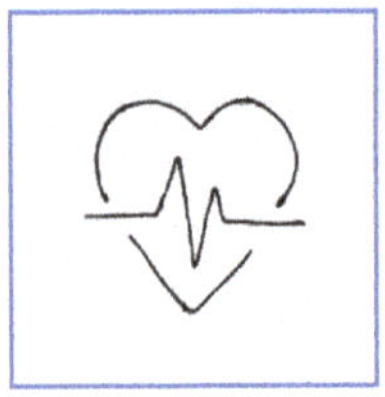

Health – The addition of green parking, bioswales, green roofs, green walls, trees, and expanded planted wells through an urban fabric improves air quality with a living green design. The trees and plants improve people's emotional health, with a living, changing green design for people to experience.

Abilities – This site design uses multiple sensory experiences. For scent, the air quality is notably cleaner, with the trees, plants, and rooftop greening. The rainwater collected from the rooftops and used for irrigation of the space can create soothing sounds for the users of the space from both nearby soothing water and the driving sounds of the urban area.

Sexuality – Luminescent lights and warm lights improve the safety of the space for womxn, the LGBTQIA+ community, and other users. Designs, signs, and street names celebrate the LGBTQIA+ community.

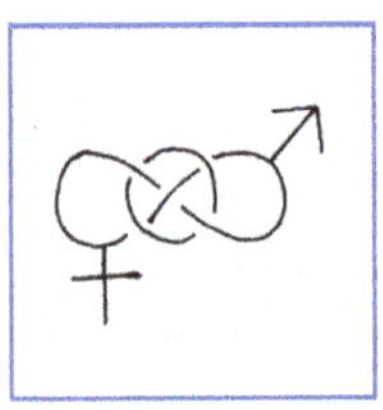

Gender – There are luminescent lights and warm lights that improve the safety of the space for womxn, the LGBTQIA+ community, and other users. Designs, signs, and street names celebrate womxn.

Age – Teenagers can have active and passive spaces where they are able to find their direction as they grow. Shared spaces for children, teenagers, adults, and mature adults to use and gather are something a community can create with uses, including seating.

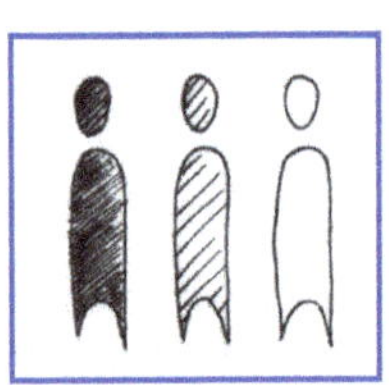

Race – In extending the applications of the SDOH, in which design of spaces is an important determinant of health, this design also applies healthy greening and other components for health equity for people who are and have been marginalized through a range of areas such as health, abilities, sexuality, gender, age, race, language, income, and culture.

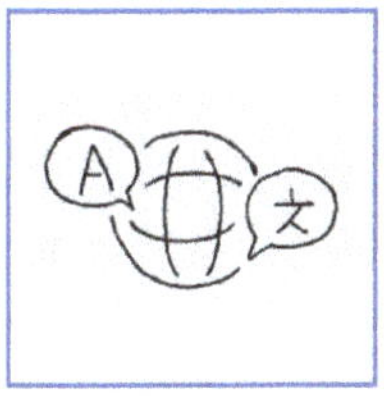

Language – Art and signs of a variety of languages and universally understood icons are designed throughout the site.

Income – Views from windows benefit users who are not easily able to experience other ways. Further, in extending the applications of the SDOH in which design of spaces is an important determinant of health, this design also applies healthy greening and other components for health equity for people who are and have been marginalized through a range of areas such as health, abilities, sexuality, gender, age, race, language, income, and culture.

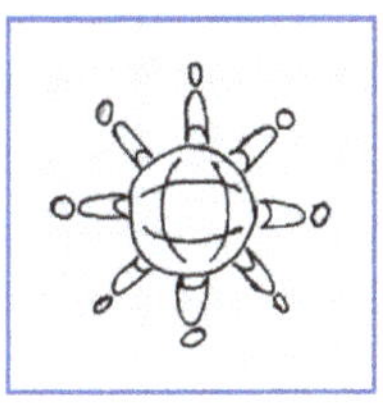

Culture – Art, designs, forms, materials, and uses of the green spaces can be culturally focused. Further, in extending the applications of the SDOH in which design of spaces is an important determinant of health, this design also applies healthy greening and other components for health equity for people who are and have been marginalized through a range of areas such as health, abilities, sexuality, gender, age, race, language, income, and culture.

Parking plazas with integrated green design of permeable pavers, bioswales, green roofs, street trees, and expanded plantings weave green infrastructure throughout an urban space, improving physical and emotional health and air quality.

Image 4.37 Green parking plaza within an urban plaza.

This design includes layers of green design ranging from pedestrian scale through large scale. The combination of permeable pavers, bioswales, green roofs, green walls, urban trees, and extended planted wells weaves green through the community, improving air quality and creating a healthy society.

Images 4.38–4.39 **Reactivating spaces under freeways and highways for public use creates new public spaces for communities that have been intentionally separated due to income and race with large obstructive physical barriers.**

4.11 Confronting the Racist Legacy of Urban Highways

Research from the book *Justice and the Interstates: The Racist Trust About Urban Highways* edited by Ryan Reft, Amanda Philips du Lucas, and Rebecca Retziaff explores the American highways' racist injustice, intentions, effects, and ideas toward reclamation (Allen, 2023).

It is not the highway itself that is racist – it is the people and agencies that located them with intention of damage that are inextricably linked to racism. Highways were built to demolish, damage, or disconnect Black communities and protect white spaces and privilege.

In concert with the Federal-Aid Highway Act of 1956 – as well as including federal, state, and local housing legislation – limited housing and economic opportunities for Blacks and Latinos, zoning, and segregation made it possible and easy to target these areas and demolish them.

Some chapters explore the lies that the politicians, planners, builders, and engineers told while developing and building the highways – that the marginalization and destruction of the Black and Latino communities were unforeseeable. Research shows that these assertions were known to be lies, and that those in power had anticipated and discussed this at the time. The practice of not having public meetings during the process of decision-makiing in these communities illustrates the intent to disregard these populations and cultures. Further, eminent domain was used in large sweeping and massive actions without acknowledging the damage to Black and Brown communities.

Black and Brown communities are still damaged and affected today by the actions of decades ago. Some impacts of the highways system include higher rates of asthma and heart disease, emotional health risks, noise pollution,

Image 4.40 **A park within an urban area woven through the highway structure allows the residents of the surrounding spaces to have recreational spaces away from vehicles.**

increased risk of premature death, neighborhood instability, and community trauma.

In Milwaukee, Wisconsin, highway construction destroyed 20,000 homes, with mostly people who were Black and Brown being displaced and jobs being lost as businesses were also demolished. Those who remained were left with a lack of public transportation, as well as poor health, due to lead poisoning and respiratory conditions from the construction and decreased air quality.

Highways create convenience for wealthy suburban areas at the expense of the health of others. Political leaders chose to place highways that avoided white neighborhoods, and through neighborhoods that lacked political power, were of people of color, and of people of low income. In a neighborhood of Montgomery, Alabama, the construction of Interstates 65 and 85 displaced 1,596 families and removed 74 small businesses, with most of these being families and businesses of color.

Reparative justice by current planners and designers is needed for these and other neighborhoods affected by highways. Actively seeking opportunities for economic revitalization, investing in infrastructure, preserving the past, affordable housing – and more – are needed.

The idea of removing the highways in Alabama, as they are seen as tributes to the American racist past, has been discussed by Professor Jocelyn

Images 4.41–4.42 **Activating community spaces with inclusive design, as well as living design, is illustrated in these play areas for children, teenagers, and adults.**

Zanzot at Auburn University. This is not a quick solution, as the automobile has taken priority in American culture.

Freeways also caused barriers to local streets, adding travel time for residents in some areas to walk or travel. One solution in the Treme neighborhood in New Orleans, Louisiana, is the activation of areas under the freeway for communities – as if defying the large intrusive freeway structure. Communities are working to claim the areas under the highways as theirs, for their use, values, and public spaces. Their long-term goal is to remove the freeway while keeping the reclaimed spaces.

Landscape architects – along with communities, planners, politicians, and more – can contribute to improve the spaces, healthy air, and social-cultural reclamation and renewal of places that have been harmed and devastated by the highway system. The U.S. can and has put funding toward the removal of highways to improve climate conditions and restore community growth.

4.12 Moving Forward with Healthy Society and Air Quality

Physical and psychological health have a reciprocal relationship that is supported by living green designs including interior greening, exterior greening, small spaces, larger spaces, creative designs, retrofitted spaces, living art, living sculptures, providing opportunities for activating spaces to improve people's health.

Healthy green designs have been notably more exclusive in areas of the United States that are of middle-class to upper-class income and predominantly Caucasian than in lower-income Black and underserved communities (Chapman, n.d.). Inclusivity in green design requires looking to retrofit interior and exterior sites with living green design to promote physical and psychological health. The layers of water access, stormwater for irrigation, greywater for irrigation, and other water sources, light sources, and sustainable materials celebrating the community and the layers of the past are specific to each site.

Listening to the community with open minds while integrating professionals with needed design skills creates a format for success for both short-term and long-term healing, health, and growth. Sites will continue to change, as will community needs, interests, knowledge, and skills. Creating spaces for sharing the changing needs and knowledge allows change to be celebrated and health to be prioritized.

Addressing equity and inclusion in green design for improving people's physical and psychological health requires addressing health, abilities, sexuality, gender, age, race, language, income, and culture in design in order to realize the effects of the past, as well as directions for immediate and long-term change. Better spaces are not the low standard proposed by the designs in this chapter. *Fully equitable and inclusive is the standard that is required* for moving forward in healthy green design and overall health.

Image 4.43 Green design woven throughout the urban fabric is a healthy approach that can be created in both public and private sites for the greater health of communities.

References

Ajzurcher. (2024). *A Greener Future for Everyone: Greenspace Inequity in San Francisco*. Retrieved 23 April 2024, from https://usfblogs.usfca.edu/sustainability/2024/04/22/a-greener-future-for-everyone-greenspace-inequality-in-san-francisco/

Allen, D. J. (2023). Confronting the Racist Legacy of Urban Highways. The Dirt, Uniting the Built and Natural Environments. *The American Society of Landscape Architects*. Retrieved 20 December 2023, from https://dirt.asla.org/2023/10/02/confronting-the-racist-legacy-of-urban-highways/

American Chemical Society. (2009, July 23). Stop and Smell the Flowers – the Scent Really Can Soothe Stress. *ScienceDaily*. Retrieved 26 April 2023, from www.sciencedaily.com/releases/2009/07/090722110901.htm

Chapman, R. (n.d.). *Parks and an Equitable Recovery*. Trust for Public Land. Retrieved 16 February 2025, from https://www.tpl.org/parks-and-an-equitable-recovery-parkscore-report#:~:text=Our%20data%20shows%20major%20disparities,in%20park%20investment%20and%20planning

Ding, L., & Hwang, J. (2020). Unequal Displacement: Gentrification, Racial Stratification, and Residential Destinations in Philadelphia. *American Journal of Sociology*, 126(2).

Evans, F. (2021, October 20; updated 2023, September 21). How Interstate Highways Gutted Communities- and Reinforced Segregation. *History, A&E Television Network*. Retrieved 20 December 2023, from https://www.history.com/news/interstatc-highway-system-infrastructure-construction-segregation

Feder, S. (2020, December 1). *Stanford Professor's Study Finds Gentrification Disproportionately Affects Minorities*. Standford Report, Standford University Social Sciences. Retrieved 25 February 2025, from https://news.stanford.edu/stories/2020/12/01/gentrification-disproportionately-affects-minorities

National Recreation and Parks Association [NRPA]. (n.d.). *Parks and Improved Mental Health and Quality of Life*. Retrieved 20 February 2025, from https://www.nrpa.org/our-work/Three-Pillars/health-and-well-being/ParksandHealth/fact-sheets/parks-improved-mental-health-quality-life/

Odom, J. (2016, November 21). Researchers Dig Deeper to Discover Healing Power of Green Spaces. *Total Landscape Care, Randall-Reilly*. Retrieved 20 December 2023, from https://www.totallandscapecare.com/business/article/15040235/the-green-road-project-helps-those-struggling-with-ptsd

Raanaas, R. K., Horgen Evensen, K., Rich, D., Siostrom, G., & Patil, G. (2011). Benefits of Indoor Plants on Capacity in an Office Setting. *Journal of Environmental Psychology*, 31(1), 99–105.

Swanson, E. (2023). Gentrification Leads to Worsened Air Quality, Study Shows. *Henry Ford Health*. Retrieved 20 December 2023, from https://www.henryford.com/blog/2023/05/gentrification-leads-to-worsened-air-quality-study-shows#:~:text=A%20variety%20of%20factors%20contribute,leads%20to%20worsened%20air%20quality

Webb, K., et al. (2024, July 31). Neighborhood Resources Associated with Psychological Trajectories and Neural Reactivity to Reward After Trauma. *AMA Psychiatry* 81(11), 1090–1100. https://doi.org/10.1001/jamapsychiatry.2024.2148

Wells, N., & Evans, G. (2003). A Room with a View Helps Rural Children Deal with Life's Stresses. *Environment and Behavior*, 35(3), 311–330.

Wolf, K. L., & Flora, K. (2010). Mental Health and Function – A Literature Review. In: *Green Cities: Good Health*. College of the Environment, University of Washington. www.greenhealth.washington.edu

World Health Organization. (2016). *Urban Green Spaces and Health*. Geneva: WHO Regional Office of Europe.

Chapter 5

Community Spaces

Images 5.1–5.4 **Shared healthy spaces are a right and basic standard for people. These can be designed to be inclusive of people with a wide range of health, abilities, sexuality, gender, age, race, language, income, and culture.**

DOI: 10.4324/9781003475293-6

5.1 Introduction

In Chapter 5, understanding community spaces includes locations that are selected for the community by an administration, spaces that the community took over ad hoc because of a site being unused or abandoned, or even those private sites shared by a community and companies.

Both interior and exterior sites are community spaces that can grow and change as the community changes. This involves being creative to retrofit unique spaces for communities that have been underserved, communities that have changing needs, and communities that originally did not have safe, green, open public spaces designed within their neighborhood for diverse people to gather together. Creating new safe healthy public spaces involves looking at unused sites, donatable sites, interior sites, underutilized sites, alleys, under highways, and other unique neighborhood opportunities for green design for a community.

In developing these community sites, both new and retrofitting existing spaces, inclusivity is important for encouraging a greater community to be heard, seen, represented, safe, and celebrated in the healthy neighborhood area's designs.

Beyond accommodating diverse people in community spaces is the practice of celebrating people. For example, creating spaces that are interesting for younger children, teenagers, and young adults to all enjoy is an asset for gathering a range of people who can teach and lead a younger generation. Similarly, creating spaces where people of mixed genders and race can be celebrated encourages people to appreciate similar and different values and knowledge.

Studies explored how race, age, and different abilities show in design, with ideas for effecting positive healthy change toward inclusive spaces for people and communities to grow. As a basic minimum standard, creating spaces which are inclusive for the full community is not just a goal, it is a basic standard in healthy design.

5.2 Industrial Warehouse Green Design

Industrialized sites have typically been zoned near low income or Black, Brown, and underserved communities in the United States (Austin Turner and Greene, n.d.; U.S. Commission on Civil Rights, n.d.; Ward, 2021). The goal of the living design in Image 5.6 is retrofitting the area around the site to be a green space to improve the health of the neighborhood and celebrate the community.

Layers of green design are integrated into the industrialized site with algae biopanels, green roofs, changing community art using air-purifying paint, lighting for safety, a variety of seating options, and use of stormwater and greywater for irrigation of the site.

The art on the site is created by the community of artists and changes with each layer of art added. Changing murals, which are spontaneously created by local and regional artists, celebrate change and the culture and race of the artists and community. Air-purifying paints are used throughout the artwork, which reduces pollutants such as nitrogen dioxide, and destroys bacteria, viruses, and mold.

Inclusivity in this design includes pedestrian-scale warm lighting throughout the space, which adds safety, particularly for womxn and the LGBTQIA+ community. Seating is included for a variety of users based on their heights and physical abilities.

In industrialized areas, the air quality is negatively affected by the pollutants produced during the processes of creating products. Green design can reduce the pollutants in the air. Algae biopanels act as living vertical art attached to the building to improve the air quality. The green roofs on the buildings have a variety of trees, grasses, and food-producing plants to add to local food sources. Plants reduce particulate matter, filtering contaminants in the air, as well as producing oxygen for people to breathe.

Irrigation for the site is through stormwater and greywater of adjacent buildings. Using stormwater in urban spaces through greening reduces the amount of water that arrives at storm sewers or storm drains, reducing the effects of flooding. Further, the green design filters stormwater for improved quality before it gets to the water table.

The equity and inclusion in this green design of urban spaces address the ideal that all people have a right to healthy shared green community spaces. In neighborhoods where green spaces have been lacking due to inequitable and exclusive design practices, creative green design opportunities must be explored at multiple layers to add to the totality of the space for improving people's physical and emotional health. The benefits of the design can be experienced by users of the site – both residents and workers, as well as the larger neighborhood.

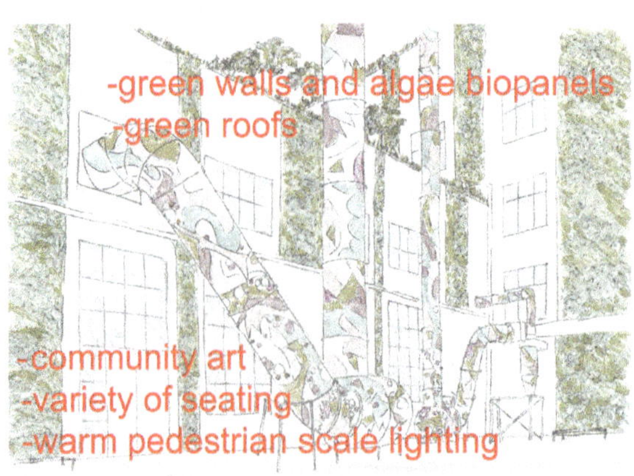

Image 5.5 Retrofitting green design at industrialized sites, both those in use and abandoned buildings, improves the health of the site as well as the community.

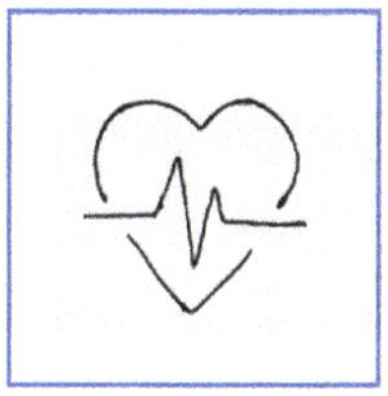

Health – The addition of artwork with air-purifying paints which reduce air contaminants, algae panels, green walls, and planted rooftops improves air quality for physical health. Emotional health is improved with the overall addition of a planted design that grows and changes form throughout time.

Abilities – This site creates a design that specifically uses multiple sensory experiences. For scent, the air quality is notably cleaner with the air-purifying paints, algae panels, and rooftop greening. The relaxing sounds include pipes capturing rainwater from the roof and decoratively running it down the side of the building.

Sexuality – Nighttime luminescent lighting as well as warm lighting improves safety. Artwork on the site is created by artists with diverse genders, ages, race, and culture.

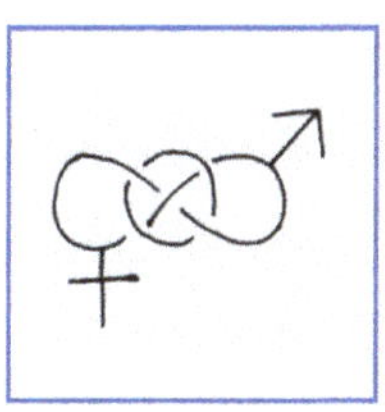

Gender – Nighttime luminescent lighting as well as warm lighting improves safety.

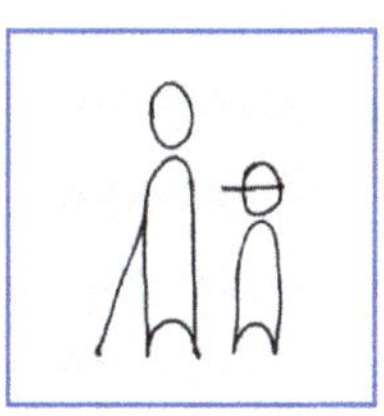

Age – The artwork on the site can change with different artists of a variety of races and cultures, time periods, and events. Keeping the site changing with plants and art allows different generations of users to create and re-create the space.

Race – The artwork on the site can change with different artists of a variety of races and cultures, time periods, and events. Keeping the site changing with plants and art allows different generations of users to create and re-create the space. Industrialized areas are typically zoned next to neighborhoods with people who have low income and are Black, Brown, and from underserved communities. Adding healthy green food plants in these industrialized public spaces can produce food for the community. This also supports people in the community who can care for, distribute, and eat the food.

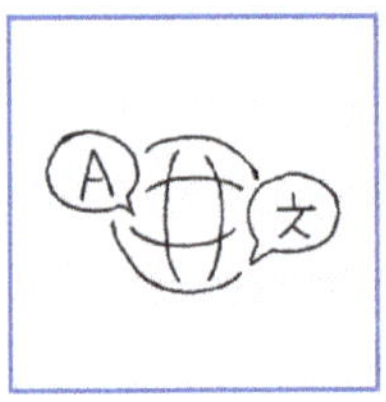

Language – Art and signs in a variety of languages and universally understood icons are designed throughout the site.

Income – Industrialized areas are typically zoned next to neighborhoods with people who have low income and are Black, Brown, and from underserved communities. Adding healthy green food plants in these industrialized public spaces can produce food for the community. This also supports people in the community who can care for, distribute, and eat the food.

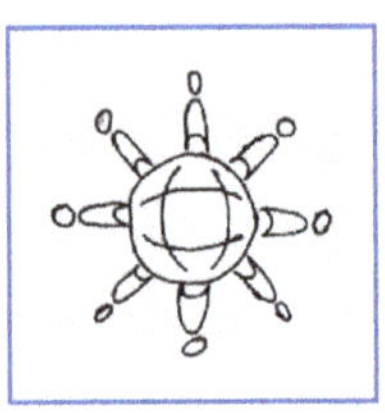

Culture – Having a greater range of plants that grow food from a variety of cultures creates a shared community experience. Additionally, the layers of artwork on the site will continue to celebrate a variety of ever-changing cultural art.

Adding green, healthy, shared spaces in industrialized and underutilized sites serve to revitalize and repurpose locales. Each design can be as unique as the neighborhood itself.

Image 5.6 **Industrialized warehouse green design.**

The exteriors of industrialized buildings are opportunities for communities to create a green space that improves the air quality, which the industry may have damaged in its production. This green site can be used to celebrate the arts while using greening for the community health.

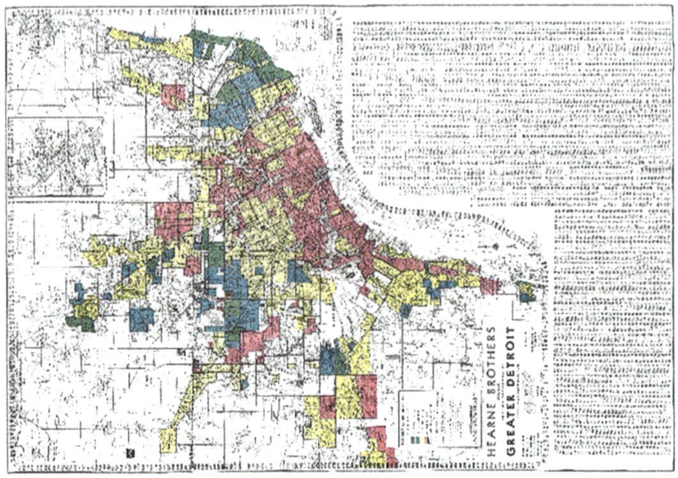

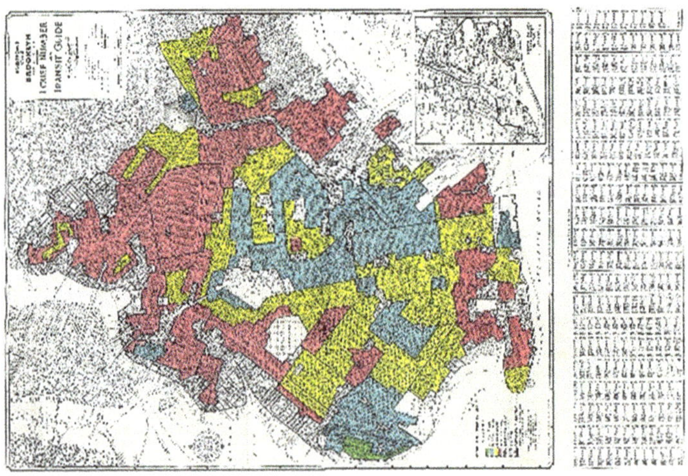

Images 5.7–5.8 **Redlined maps of Detroit, Michigan, and Brooklyn, New York, in the United States illustrate how communities were mapped to separate people based on race.**

5.3 Kofi Boone: Designing a Black Commons

Kofi Boone spoke of creating a Black Commons by pooling collective land and resources to stabilize and empower Black communities (Green, 2023). With the effects of racial and structural inequalities, the effects of a Black Commons are to generate wealth in underserved communities. "After experiencing decades of redlining, urban renewal, gentrification, and displacement, Black communities can combat systemic issues by envisioning new communities that are mutually supportive."

Images 5.7 and 5.8 illustrate maps of Detroit, Michigan, and Brooklyn, New York, in the 20th century where areas of the map were coded by color. The federal government enabled these maps to be made by the Homeowners Loan Corporation across the United States. Areas marked in red would receive no loans from the federal government or banks. The redlined communities were predominately Black communities populated by families unable to create or grow generational wealth.

Image 5.9 **Without design solutions, houses on the low part of the slope flood.**

Further, these redlined areas were not included in federal programs to add street trees, public comforts, and significant infrastructure. The effects of these include a lack of deep shade canopies creating hotter areas, poor air quality, poor health, and a lack of space for people to gather in their community (Boone, 2023).

Boone's (2023) research and projects to advance Black Commons include the following.

The Bennehan and Cameron families' 30,000-acre plantation in North Carolina once had 1,000 people who were enslaved working on the farm. The enslaved people involuntarily invested in the generational wealth of these families while being unable to invest in their own. Working with Urban Community Agrinomics (UCAN), North Carolina State University is developing a vision for wealth generation through a Black-owned cooperative.

Another project includes preserving and restoring a college prep school and campus for Black students in the South. Boone states that preserving sites of cultural significance is important, saying, "If we don't value these stories, then we can't continue telling them."

Another project is the Freedom Hill site in Princeville, North Carolina. It was common for Caucasians to live on high ground while Blacks lived in lower areas, typically closer to the plantations they worked. Presently, this area is still a Black community on low land that regularly floods. The U.S. Federal Emergency Management Agency (FEMA) came to assist the community, but, rather, they overloaded them in ways that caused more problems. Andy Fox, a professor at N.C. State, developed a guide for the community to better understand their options and develop design-based solutions.

Boone's project highlights creating Black Commons with shared land ownership and administration and support of wealth generation, health, and empowering the culture.

Image 5.10 **The effects of flooding are devasting to families and entire communities.**

5.4 Urban Park Under Highway Green Design

Roadways, freeways, and highways have been used to separate neighborhoods to create exclusivity and inequity within a city and region (Evans, 2023). Further, the roadways add air pollutants to neighborhoods, disproportionately to areas with low income and Black, Brown, and underserved communities (Austin Turner and Greene, n.d.; U.S. Commission on Civil Rights, n.d.; Ward, 2021). The design in Image 5.12 reconnects neighborhoods and communities by weaving safe green parks and community spaces under highways for pedestrians.

The green design reuses the public space as a green park, irrigated by the stormwater and greywater produced by the adjacent buildings, as well as adding light for safety, and reduces sound with the use of plants. A variety of seating options are included throughout the space. The buildings and surfaces in the neighborhood also have green roofs, green walls, and algae biopanels. Irrigation for the site is through stormwater and greywater of adjacent buildings. Using stormwater in urban spaces reduces the overworked public water systems.

The redesign of this space creates a pedestrian greenbelt weaved through the urban area, which improves the physical and emotional health of the community with a park system designed for people to walk to and through to other neighborhoods. This green design improves air quality, including reducing particulate matter, adds oxygen to the air, and reduces sound waves.

Inclusivity in this design also includes lighting under the freeways in a constellation pattern, as well as pedestrian-scale warm lighting throughout the space, which adds safety, particularly for womxn and the LGBTQIA+ community. Seating is included for a variety of users based on their heights and physical abilities.

Creating ways to integrate neighborhoods that have been separated serves a greater community toward inclusion. Equity with green design addresses the ideal that there is a right for all people to have healthy green community spaces. This requires retrofitting green design to weave communities together, improving people's physical and emotional health.

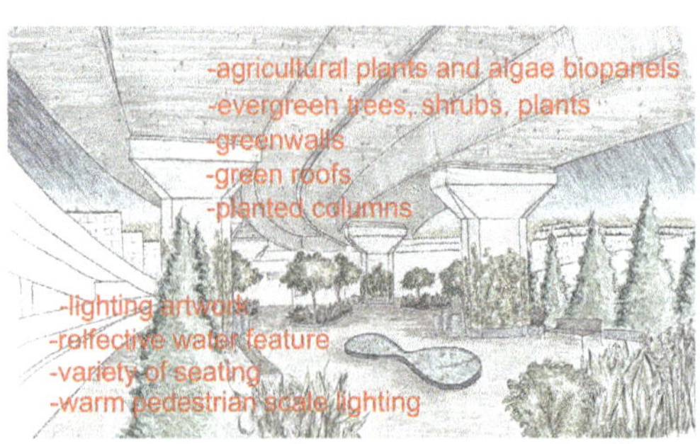

Image 5.11 **Retrofitting green areas into communities adjacent to roadways, freeways, and highways improves the emotional and physical health of the people of the community.**

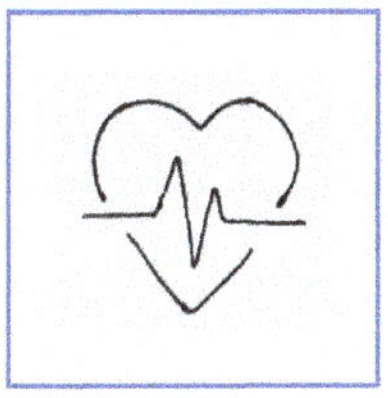

Health – The addition of green parks under freeways and highways improves air quality with trees and plants as well as emotional health with more green spaces for spending time. There are luminescent lights and warm lights, which appear to be stars, under the freeway, improving the safety of the space for womxn, the LGBTQIA+ community, and other users.

Abilities – This site design specifically uses multiple sensory experiences. For scent, the air quality is notably cleaner with the trees, plants, and rooftop greening. The rainwater collected from the nearby area and run throughout the park can create soothing sounds for the users of the space to hear both nearby soothing water and driving sounds of the urban area.

Sexuality – Luminescent lights and warm lights, which appear to be stars under the freeway, improve the safety of the space for womxn, the LGBTQIA+ community, and other users.

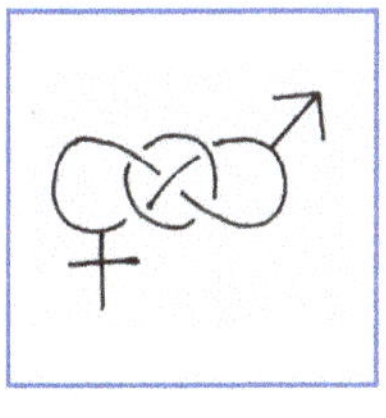

Gender – There are luminescent lights and warm lights, which appear to be stars under the freeway, improving the safety of the space for womxn, the LGBTQIA+ community, and other users.

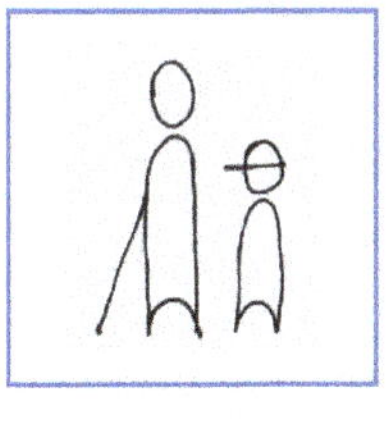

Age – Shared spaces for teenagers to use and gather are something a community can create in these unique green spaces. Teenagers can have active spaces or passive spaces where they are able to find their direction as they grow.

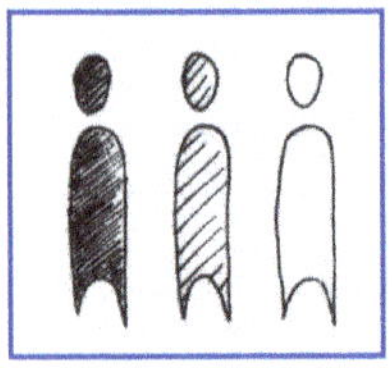

Race – Highways and freeways are typically zoned next to neighborhoods with people who have low incomes and are Black, Brown, and underserved communities. The roadways create designs that separate neighborhoods. Adding green designs, parks, and playgrounds under the roadways creates healthy pedestrian areas to connect communities, improve air quality, and reduce vehicular noise.

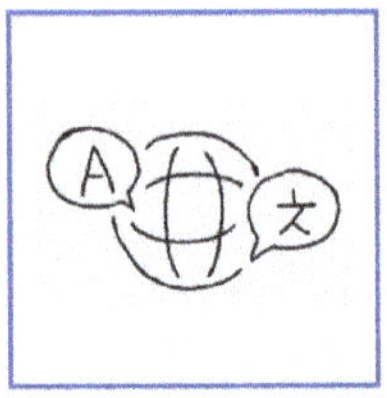

Language – Art and signs in a variety of languages and universally understood icons are designed throughout the site.

Income – Highways and freeways are typically zoned next to neighborhoods with people who have low incomes and are Black, Brown, and underserved communities. The roadways create designs that separate neighborhoods. Adding green designs, parks, and playgrounds under the roadways creates healthy pedestrian areas to connect communities, improve air quality, and reduce vehicular noise.

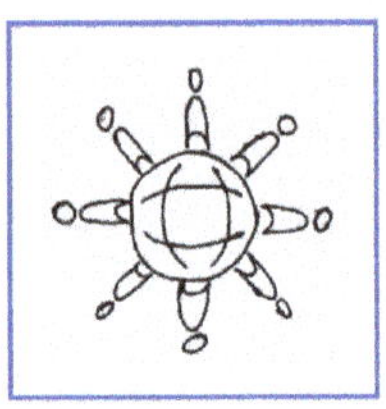

Culture – Designs, forms, materials, and uses of the park can be cultural focused. Highways and freeways are typically zoned next to neighborhoods with people who have low incomes and are Black, Brown, and underserved communities. The roadways create designs that separate neighborhoods. Adding green designs, parks, and playgrounds under the roadways creates healthy pedestrian areas to connect communities, improve air quality, and reduce vehicular noise.

The benefits of greenbelts include connecting communities, improved air quality for the health of people and animals, and green spaces for animal habitats.

Image 5.12 **Urban park under a highway.**

The green design of the highways, freeways, and roadways that once were used to separate people is now used to connect neighborhoods and communities. Creating safe greenbelts under freeways encourages people to connect and celebrate their spaces, sharing green resources for a greater common good.

Image 5.13 **Portland Memory Garden in Portland, Oregon, is designed for people with Alzheimer's, as well as for their family and caregivers.**

5.5 Ed Benedict Park and Portland Memory Garden

The Ed Benedict Park and Portland Memory Garden have two notable pieces in reference to this chapter on community spaces (Bainnson, n.d.).

The first is that the space of the community park was slated to be used for a freeway. However, neighborhood activists stopped this area from having a freeway through it and were successful in having the space designed as a community park. With a notable amount of the community in this area living

Image 5.14 **Designing with multiple sensory experiences encourages memory use.**

Image 5.15 **Vibrant colors, a variety of forms and textures, fresh scent, and sounds of wind through plants create spaces of emotional therapy in this garden and park.**

in building complexes, the public community green spaces are what the community wanted.

A second notable feature of this park is that it includes a quiet and tranquil area designed for those people with Alzheimer's disease and their caregivers. This allows time and space for people to decompress, relax, and look inward for peace and clarity.

The Ed Benedict Park and Portland Memory Garden includes features that are expected in modern parks, such as providing places for children to play and for people to walk, run, garden, grow organic food, and choose active

Image 5.16 **Walks that are safe, as well as having engaging views and seating, provide a green space for people with memory issues to be intentionally included and visible in public spaces.**

recreation including sports. Exercise in nature has physical and emotional health benefits for the users of the park.

Additionally, there is a memory garden, a special garden designed for people with Alzheimer's or other types of memory loss. In 2020, it was estimated that more than 40,000 people in Portland have Alzheimer's disease, and over 75% of them live at home (Alzheimer's Association, n.d.; Bainnon, n.d.). With this information, the city determined that this type of sensory garden is a required service to the community. The Memory Garden has therapeutic effects for people with diseases of memory loss, as well as for their caregivers.

The design of the Portland Memory Garden uses specific plants laid out to support the reduced abilities of those with memory loss, as well as designs that can help stimulate pleasant memories. The design of this garden within the park encourages care and growth for a larger inclusive community. With people of different ages and abilities able to use the public space, the community has designed considerate accommodations for a greater inclusion of the population.

5.6 Gymnasium Garden at Public School Green Design

Public schools in the United States are underfunded (Zdanowicz and Yan, 2022). Further, the demographics of the race and income of students in public schools (relative to those in private schools) in the United States is disproportionate, with more students having low income and being Black,

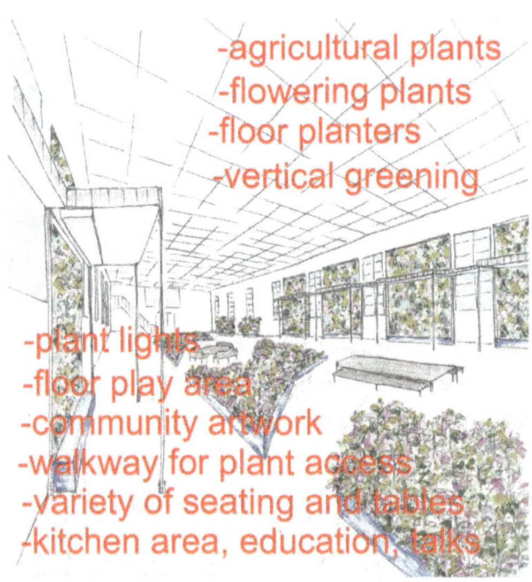

Image 5.17 **Greening an interior in a public school includes agricultural, flower-producing, and other interior plants designed throughout the garden.**

Brown, and from underserved communities (National Center for Education Statistics, 2024). The design in Image 5.18 serves as layers of educational growth, as well as physical and emotional health for the students, faculty, and staff of the school, as well as the families of these people and the community of the school.

The indoor gymnasium garden is designed with planters at a variety of horizontal and vertical heights. Plants are agricultural, as well as flowering, and other plants are designed throughout the indoor space. Plant lights are added under the platforms for the living green walls, as well as additional plant lights throughout the space. There is a pulley system to lower plants down from the platform.

An area for teaching, as well as a sink and food preparation area, are designed within the space for teaching about agriculture, growing food, healthy food, and preparing food. A variety of seating and tables are included for users based on their height and physical abilities. The walkways are comfortable widths for children walking with adults, as well as for people with different walking abilities.

Areas for play, art, presentation, and talks benefit children and people of all ages physically and emotionally. Being exposed to healthy growing green interior designs can encourage people of different health, abilities, sexualities, gender, age, race, languages, incomes, and culture to work together as a community while breathing in fresh air and enjoying their unique garden. Interior green design within a school system celebrates equity and inclusion as part of the educational system. When learning these practices at an early age, children are able to weave this into their own practices and styles, creating a new generation of equity and diversity.

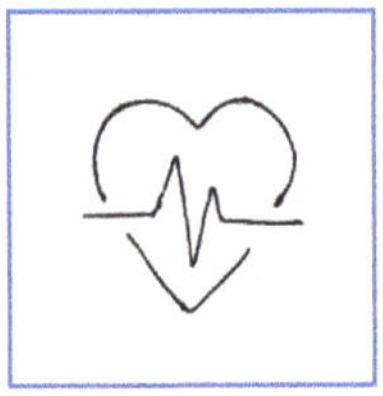

Health – This indoor green space creates a year-round opportunity for the community to grow and eat locally grown food and sell floral displays. The physical health from the improved air quality in this site, as well as the emotional health of a planted design that grows and changes is beneficial to the users.

Abilities – Creating walking paths with textures, colors, signage, and icons helps those with different visual abilities to navigate spaces. Scent in this space also helps users with different visual abilities to understand their surroundings. Further, this site opens opportunities for people with different learning abilities to have quiet spaces with reduced sounds to be able to focus and learn with fewer distractions.

Sexuality – Signage of local businesses welcomes and celebrates people of the LGBTQIA+ community.

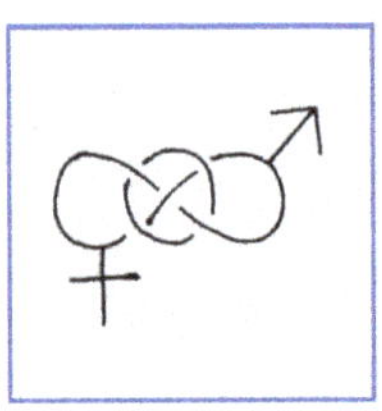

Gender – Creating a source for local healthy food for womxn who work at home is a considerate choice for their time and health. Designs and signs celebrate womxn.

Age – This indoor green space encourages community members ranging from early schooling through mature adults to enjoy the space together and support the agriculture. The design includes low planters and higher planters for users of different heights, including users in wheelchairs, to enjoy the planted space.

Race – Creating a year-round food source in a public school supports the students, faculty, staff, families, and community. The artwork on the site can change with different artists of a variety of races and cultures, time periods, and events. Keeping the site changing with plants and art allows different generations of users to create and re-create the space. Further, in extending the applications of the SDOH in which design of spaces is an important determinant of health, this design also applies healthy greening and other components for health equity for people who are and have been marginalized through a range of areas such as health, abilities, sexuality, gender, age, race, language, income, and culture.

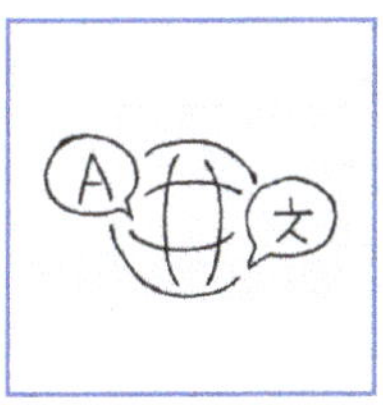

Language – Art and signs in a variety of languages and universally understood icons are designed throughout the site.

Income – Interior community spaces that produce food and flowering plants are beneficial to the community. The students in some public schools include families with low income. Their need for food security is addressed with an indoor green space. Healthy, locally grown food can benefit people with low income who need the food, as well as people in the community who can care for and distribute the food.

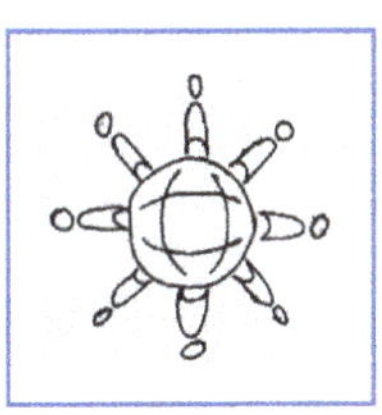

Culture – Creating designs with plants of a variety of cultures benefits the community to be educated on diverse cultures of the community and world. Indoor community spaces that produce food and flowering plants are beneficial to communities. Healthy, locally grown food can benefit people as well as support people in the community who can care for and distribute the food.

Redesigning public and private school gymnasiums to have public green spaces for the community encourages people with diverse abilities, sexualities, genders, ages, races, ages, skills, incomes, and cultures to share a space together and learn from the community. The food produced in this year-round space can also be used for community meals throughout theyear.

Image 5.18 Green gymnasium garden at a public school.

Turning a gymnasium into a school and community space is for the education of the children and community. This site provides education, food, and flowers for sale, and provides a healthy space for a tired soul to relax and refocus when needed.

Image 5.19 **Topiary garden at the Old Deaf School Park in Columbus, Ohio.**

5.7 Two Unique Community Gardens

5.7.1 Topiary Garden at Old Deaf School Park, Columbus, Ohio, United States

Located in the Old Deaf School Park, this evergreen sculptured design is entirely based on the painting by George Seurat, *Sunday Afternoon on the Island of Grande Jatte.* This garden's name celebrates the school, campus, and deaf culture that once was prominent on the 9.2-acre site. The oversized

Image 5.20 **The evergreen topiaries are the figures from the painting *Sunday Afternoon on the Island of Grande Jatte* by George Seurat.**

Image 5.21 **The unique community park is located on 9.2 acres of the former campus of the Ohio School for the Deaf.**

figures represent those depicted in the 1884 painting and are maintained by the Columbus Recreation and Parks Department.

The uniqueness of the park itself is notable. However, it is the name of the park that is important for acknowledging, celebrating, and including a culture of people. Deaf and the hearing impaired are a culture of people, not a language. This is because people who are deaf share a unique set of experiences, traditions, and language, which set them apart from the hearing community.

For example, in learning the visual American Sign Language (ASL), there are multiple differences compared with the spoken English language. ASL is not translated on a word-by-word grammatical basis as English grammar. The order of the words in a sentence is not grammatically the same as spoken English. "I went to the store yesterday" in spoken English is visually gestured in ASL as "Yesterday store I go," with the tense being first. Past and present tense are relayed differently, and some ASL words or gestures can only be understood in context of the sentence or topic.

Image 5.22 **The evergreen design acknowledges and celebrates the deaf culture in the name of the park.**

5.8 Els Center for Excellence, Sensory Arts Garden

Another unique community park is in Jupiter, Florida: the Sensory Art Garden and Pavilion designed for people on the autism spectrum (Els Center for Autism, n.d.). The goal for the 13,000-square-foot space is to expand art integration for people of varying ages and abilities by designing and programming a creative place that can be experienced via touch, sight, hearing, smell, taste, and motion.

The garden has lush green rooms with accessible planters, seating, a water feature, and selected plants, offering an intimate and calming sensory experience for a group or an individual. These spaces stimulate motor, cognitive, sensory, and social skills.

The garden is designed with varying levels of sensory experiences, while minimizing sensory overload. This is both therapeutic and educational, especially for those with sensory processing challenges including autism and other related different abilities. Engaging with nature helps people with autism to develop interests in birds, animals, plants, and social skills.

5.9 Interior Lobby Community Room Green Design

There are a large variety of interior spaces to which the design in Image 5.24 of a vertical living green art mural can be applied. Even in rooms where floor space is lacking or is required for other uses, a full green wall creates a stunning green space for people to share.

Plant lights are placed along the wall of plants, and a variety of seating for people with different abilities is designed into the space. Fresh cut flowers can be put in some containers and changed out for a fun splash of color and scent as wanted.

The vertical living green mural uses plant cuttings grown in water and hung on a wall structure. Roots of the plants are visible in clear containers, allowing people to see the roots grow and change, as well as the green structure of the plant growing above the waterline. As the plants grow too large for the container, the plant can be changed to a larger container or be cut and propagated, with the cuttings to be shared with the community.

Giving away plant cutting is an approach to inclusivity, education, and health. Sharing plants throughout a community brings people together to share stories and also greens up a larger community. As plants continue to grow, more plants can be shared, creating a healthier space for more and more people. The air quality is improved with the growing number of plants, and people share time, knowledge, and shared experiences.

Giving away plant cuttings and larger plants to people no matter their health, ability, sexuality, gender, age, race, language, income, or culture is an equitable approach to adding green design to people who may not have as much green community space in their neighborhood. The ideal that health is a right of all people is practiced in this design as fresh air is improved, which is beneficial for people both physically and emotionally.

Image 5.23 Interior green design creates spaces for people to gather in any weather. This design uses an entire wall to create a living artwork that grows and changes to be enjoyed on a daily basis.

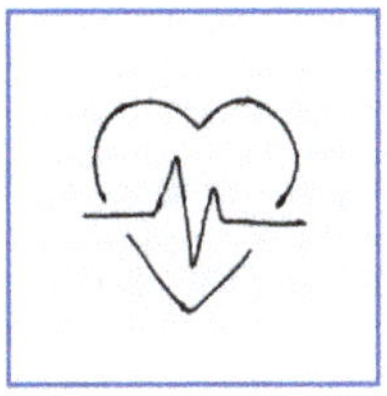

Health – The addition of year-round interior green sites encourages people to gather in warm, healthy green spaces during the winter. Spending time in a green space during cold months improves emotional health, as well as physical health, with healthy air quality.

Abilities – The indoor community green space has plants of different heights for direct view in people's eyeline. The site also changes as plants are propagated, grow, and are cut to further propagate. The changing spaces encourage people to experience healthy green visible change over a short time. With change and other sensory design components, emotional health is improved by seeing beautiful change in other living forms and relating to the ability to change and grow.

Sexuality – Signage of local businesses welcomes and celebrates people of the LGBTQIA+ community.

Gender – Designs and signs celebrate womxn in the community.

Age – This indoor green space encourages community members ranging from early schooling through mature adults to enjoy the space together. The design includes low planters and higher planters for users of different heights, including users in wheelchairs, to enjoy the planted space.

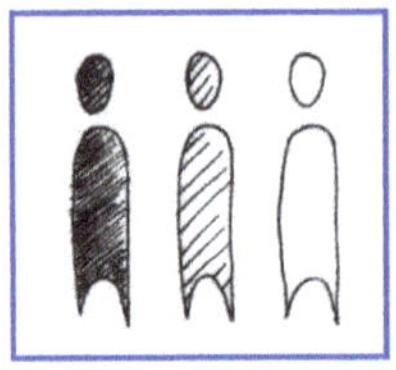

Race – In extending the applications of the SDOH, in which design of spaces is an important determinant of health, this design also applies healthy greening and other components for health equity for people who are and have been marginalized through a range of areas such as health, abilities, sexuality, gender, age, race, language, income, and culture.

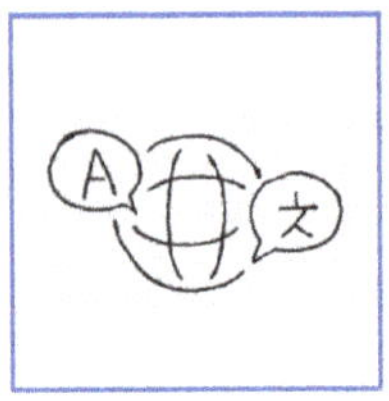

Language – Art and signs in a variety of languages and universally understood icons are designed throughout the site.

Income – As the plants on the wall garden grow, they can be cut, propagated, and shared with people throughout the community. Sharing plants allows people to put more living designs in their homes, creating healthier air and spaces throughout a community. Giving plants to people who cannot afford them is part of spreading health through the community.

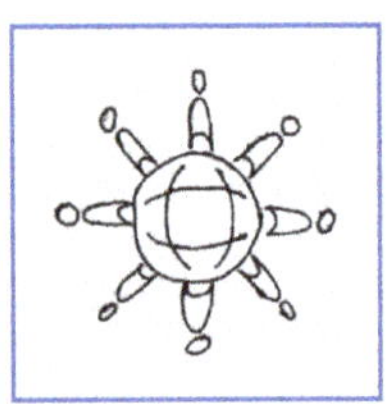

Culture – Indoor community spaces that have plants and designs from a variety of cultures are beneficial to the community, who can learn and celebrate the great range of cultural diversity in a green space. As the wall grows, cuttings of the plants can be shared throughout the community with individuals as well as for greening more sites.

Adding year-round interior green spaces encourages people to gather as a community. This style of design for an interior can be used in a lobby, office, meditation room, airport community spaces, or a variety of other types of interior year-round spaces.

Image 5.24 **Interior green lobby or room for the public.**

This year-round interior site becomes a community celebration of growth and changes in a warm space. The interior site has plants growing in containers with water. The visibility of the roots in the water is educational, as well as relaxing, to see the growth of the plant in both the leaves and the roots. As the plants grow beyond the size of the container, cuttings from the plants can be shared with the community.

5.10 Moving Forward with Healthy Green Community Spaces

The importance of shared green community spaces cannot be overstated. In order to grow one's own sense of self, share time and dialogue with others, understand the world around us, learn our direction and voices, protest inequity, celebrate differences, and find ways to work together and understand the views, needs, wants, and interests of others, we need safe healthy spaces to achieve these goals. Some communities have more green open spaces accessible to them, while others do not.

Creating exterior and interior spaces for communities to gather requires creativity to retrofi; an inclusive approach is looking at both smaller and larger spaces that are accessible by walking and public transportation, are safe, are inclusive for people with different abilities and needs, and allow people to come together. The spaces can vary in scale and have a variety of short-term and long-term changes.

For creating new spaces, starting at a point where green design is required in the planning of the community for people to gather, walk, play, work, commute, talk, share ideas, celebrate events, experience theater and other arts, and so much more is an ideal that is attainable. As the health, abilities, sexuality, gender, age, race, language, income, and cultures of a great variety of people are listened to, understood, practiced, and applied through green design, the larger community grows and expands.

Equity and inclusion as an approach to community green design enriches and raises both individuals and a larger community. Through adding green community spaces, individuals and the community grow, expand, create, and rise. This ideal is attainable by retrofitting green designs in existing communities, as well as creating new green communities with green design.

Image 5.25 Community spaces, including those that private business owners create to share with the community, are spaces for celebrating the rich and growing diversity of people. Making sure all voices are heard, listened to, and celebrated reminds the larger culture that equity and inclusion are assets in growth and health.

References

Alzheimer's Association. (n.d.). *Oregon and SW Washington Chapter*. Retrieved 6 February 2025, from https://www.alz.org/orswwa#:~:text=In%20Oregon%2C%20more%20than%2079%2C000,million%20hours%20of%20unpaid%20care

Austin Turner, M., & Greene, S. (n.d.). Urban Institute. Retrieved 16 February 2025, from https://www.urban.org/racial-equity-analytics-lab/structural-racism-explainer-collection/causes-and-consequences-separate-and-unequal-neighborhoods

Bainnson, B. (n.d.). Ed Benedict Park and Portland Memory Garden. *American Society of Landscape Architects*. Retrieved 20 December 2023, from https://www.asla.org/portland/site.aspx?id=43611

Boone, K. (2023, February 15). *Recognition, Reconciliation, Reparation* [Lecture]. Harvard Graduate School of Design. Retrieved from https://www.gsd.harvard.edu/event/kofi-boone-recognition-reconciliation-reparation/

Els for Autism. (n.d.). Retrieved 20 December 2023, from https://www.elsforautism.org/the-els-center-of-excellence/sensory-arts-garden/

Evans, F. (2021, October 20; updated 2023, September 21). How Interstate Highways Gutted Communities- and Reinforced Segregation. *History, A&E Television Network*. Retrieved 20 December 2023, from https://www.history.com/news/interstate-highway-system-infrastructure-construction-segregation

Green, J. (2023). Kofi Boone: Designing a Black Commons. The Dirt, Uniting the Built and Natural Environments. *The American Society of Landscape Architects*. Retrieved 20 December 2023, from https://dirt.asla.org/2023/02/21/kofi-boone-designing-a-black-commons/

National Center for Education Statistics. (2024). *Concentration of Public School Students Eligible for Free or Reduced-Price Lunch*. Condition of Education. U.S. Department of Education, Institute of Education Sciences. Retrieved 16 February 2025, from https://nces.ed.gov/programs/coe/indicator/clb

U.S. Commission on Civil Rights. (n.d.). *Not in My Backyard: Executive Order 12,898 and Title VI as Tools for Achieving Environmental Justice*. Retrieved 16 February 2025, from https://www.usccr.gov/files/pubs/envjust/ch2.htm

Ward, Jr., K. (2021, December 21). How Black Communities Become "Sacrifice Zones" for Industrial Air Pollution. *ProPublica*. Retrieved 16 February 2025, from https://www.propublica.org/article/how-black-communities-become-sacrifice-zones-for-industrial-air-pollution

Zdanowicz, C., & Yan, H. (2022, September 18). US Public Schools Get a D+ for Poor conditions, and Experts Say Problems are Getting Worse. Here's What Kids are Facing. *CNN*. Retrieved 16 February 2025, from https://www.cnn.com/2022/09/18/us/school-conditions-2022/index.html

Chapter 6

Access to Public Comforts and Safety

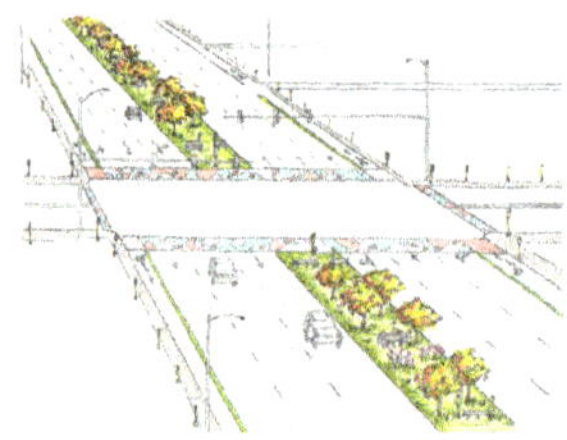

Images 6.1–6.5 **Public spaces small scale to large scale where people spend short amounts of time, as well as those where people engage in the space for an extended amount of time have inclusivity as a primary goal to create comfort for their users.**

DOI: 10.4324/9781003475293-7

6.1 Introduction

This green design chapter addresses designs of areas for public comforts, as well as designs addressing safety for people from marginalized communities. Spaces for public comforts have a few directions that need to be focused on for inclusive design, such as designs that address health, ability, sexuality, gender, age, race, language, income, and culture. As an example, in areas with people who have a low income, there are fewer spaces for public gatherings and notable green spaces for the health of the people in the community.

> [D]ata shows major disparities in access to the outdoors. In the 100 most populated cities, neighborhoods where most residents identify as Black, Hispanic and Latinx, American Indian/Alaska Native or Asian American and Pacific Islander have access to an average of 44 percent less park acreage than predominantly white neighborhoods, and similar park space inequities exist in low-income neighborhoods across cities, highlighting the urgent need to center equity in park investment and planning.
>
> (Chapman, n.d.)

People experience a sense of safety differently and have a variety of needs for safety in public spaces. Safety involves not only bright lighting. Safety is also about the color and quality of the lighting, the distance the light carries (luminance), the views into the spaces, subspaces for semi-privacy, materials of the walkway to walk safely, the sounds in the space, the context of the space, and much more.

There are a few studies in this chapter that explore safety and how womxn and people who are LGBTQIA+ experience safety differently, and how marginalized populations want public spaces to address safety for all. Safety includes lighting materials, luminance, subspaces, views into and out of the space, sounds in the spaces, context of the space, pedestrian scale, speed of traffic, types of transportation available, uses of the space, ability to safely navigate a space solo, ability to navigate a space while with a child or children in one or more strollers, ability to navigate a space while with someone using a walker, ability to use the space as needs change for people, and other safety needs.

When designers create spaces for people who are marginalized based on their health, abilities, sexuality, gender, age, race, language, income, and culture, the spaces become inclusive and safer, and they encourage greater use of the area. These goals are important for communities to grow in both the short term and the long term.

6.2 Expanded Sidewalks for Shared Space and Safety, Green Design

Image 6.7 includes wider sidewalks, allowing a safe space for different uses of the space for running, walking, strolling, passing, shopping, and sitting, and space for expanded tree wells and planters, trash, postal dropoffs, streetlights, and more. The totality of the green interventions improves the emotional health of people, as well as the air quality. Wider sidewalks focus on the safety and comfort of pedestrians and are designed for equity and inclusivity through layers of details and interventions.

Business zones include window displays, signage, entries, seating, and outdoor sales, as well as greening with planters and green roofs for the emotional and physical health of the community. There is a wide zone for pedestrians for running, wheeling, walking, using strollers, or walking with people at different paces. There is a zone for features such as tree wells, which are expanded for further plantings and seating, trash disposal, streetlights, postal dropoffs, bus stops, bicycle parking structures, street names, community signage to celebrate events and people, public art, and other similar features. This same zone also has a curb cut for safely crossing the street with colorful rumble strips. Beyond the sidewalk, there is a protected bicycle lane for cyclists and scooters, and then vehicle lanes.

Image 6.6 Creating sidewalk zones for different users encourages safer use for people. With safety for a variety of users to share the space, they are able to also enjoy the space as they walk through the neighborhood.

The wide sidewalks with zones create inclusivity through a variety of safe and considerate designs for a variety of users. These include pedestrian-scale warm lighting, which addresses safety. Womxn and people of the LGBTQIA+ community identify safety in design as a want and need to feel comfortable engaging with sites and feeling included in their community (Boomsma and Steg, 2014; Cattarall and Azzouz, 2021; Chaney et al., 2024; Etopio et al., 2019; Peachey and Baller, 2015).

Further, signage with symbols and diagrams, as well as multiple languages, is put at a variety of heights for people who use wheelchairs and children to read at their height. Designs for inclusivity of children also include scented and colorful plants at lower heights, as well as seating for a variety of users for the bus stops, expanded tree well gardens, and business seating.

Wide pedestrian walkways are important for equity and inclusivity. Inclusivity for a larger community is important on both a small scale and large scale when addressing public spaces. Listening to how people experience space differently creates most inclusive factors needed when designing public comforts and shared spaces.

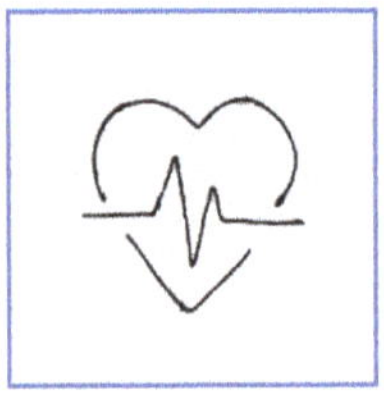

Health – The addition of expanded tree wells and planters improves air quality for easier breathing. The addition of green design improves people's emotional health. Pedestrian safety is improved with zones for moving as well as a protective planted edge blocking vehicles.

Abilities – Zones for pedestrians, curb cuts, detectable rumble strips and warning pavers at corners, crosswalk melodies for walking and do not walk, and high-contrast signage improve navigation for people with different visual and walking abilities. A variety of seating options and spaces for people with different mobility skills are available.

Sexuality – Consistent luminescent lighting at pedestrian scale along each block, as well as warm lighting, improves safety for the LGBTQIA+ community. Business signage, designs, street names, business merchandise, and services welcome and celebrate people of this community.

Gender – Consistent luminescent lighting at pedestrian scale along each block, as well as warm lighting, improves safety for womxn in the community. Signage of local businesses welcomes and celebrates people of this community. Caregivers of children and senior adults have wider space for strollers and holding children's hands. Business signage, designs, street names, business merchandise, and services welcome and celebrate womxn.

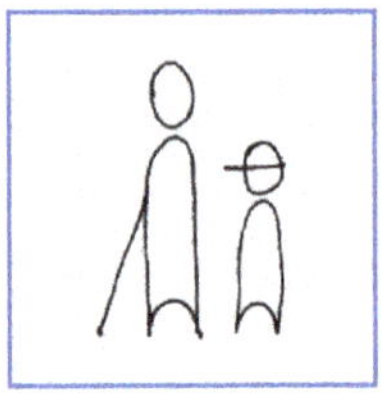

Age – Curb cuts and safe crosswalks are at every street corner for people to safely cross streets at multiple locations on their route. This is important for people who walk throughout their day while running errands with children and people who are seniors. Fun designs are placed at lower heights for the sightlines and tactile experience of children.

Race – Business signage, designs, street names, business merchandise, and services welcome and celebrate people who are Black, Brown, and from underserved communities.

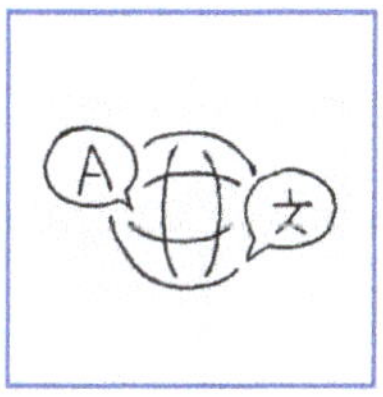

Language – Signs that use multiple local, regional, and international languages, as well as icons and images without words for those learning language, are throughout the street at pedestrian heights.

Income – Food-producing plants, which are designed throughout the streetscape, are available to eat to support people with low income in the community who want or need healthy food. The food can also be distributed locally to people or foodbanks.

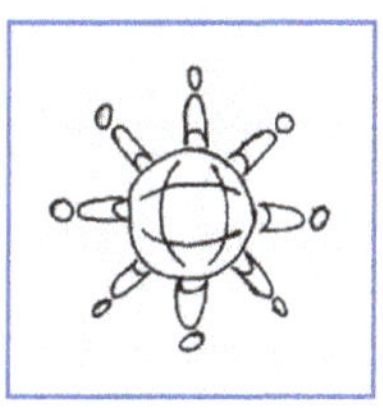

Culture – Business signage, designs, street names, business merchandise, and services welcome and celebrate people of different local, regional, national, and international communities.

Adding sidewalk zones for equity and inclusion further creates a pedestrian scale site with improved safety and public comforts.

Image 6.7 **Sidewalk zones in expanded sidewalk design for pedestrian scale and safety.**

Illustrate sidewalk zones for improving safety of the users with areas for: a.) business display windows and store entry; b.) area for tables, signs, and greening; c.) area for walking; d.) area for lights, trash, postal boxes, expanded tree wells with planters, safe curbs for crossing, street signs, and bike and scooter parking; e.) curb for safety of pedestrians near driving lanes; f.) driving lanes; g.) bike lane (off image after two lanes of cars on the other side of the street).

Image 6.8 Manchester's Gay Village celebrates an underserved population with murals, signs, and safe streets.

6.3 The Queer City: How to Design More Inclusive Public Spaces

Pippa Catterall and Ammar Azzouz (Catterall and Azzouz, 2021) write of how spaces are not designed for marginalized communities, including people who identify as LGBTQIA+. In a myriad of countries, the architecture profession is dominated by heterosexual Caucasian men who create spaces from that perspective. This includes male drinking spaces and parks with sports facilities built primarily for boys.

Image 6.9 London's Soho celebrates the LGBTQIA+ culture with safe streets and public displays of inclusive art and design.

Images 6.10–6.11 **Vienna, Austria, uses gender-sensitive lighting for safe and inclusive design in Image 6.10. Image 6.11 in Klapparstigur, Iceland, celebrates the LGBTQIA+ community as inclusive of safe accessible design for the community.**

This affects how people who do not follow the antiquated masculine stereotype, including womxn, experience these spaces. With the addition of poor lighting, marginalized groups are made to feel uncomfortable and therefore do not feel welcome or go to these spaces.

Since the 1980s, research has been done on mapping public spaces for how marginalized communities access and use the spaces. One approach to improving spaces for inclusion of LGBTQIA+ is called "queering" public spaces.

In areas such as Manchester's Gay Village and London's Soho, "queering" these public spaces began in run-down, marginal areas of the city. These areas had low rent and access to public services. With people who are LGBTQIA+ inhabiting these spaces and building up the space, the areas improved and became a safer space for their inhabitants. However, as the area improved, the rents rose exorbitantly high, and it became overrun with wealthy heterosexual Caucasian men. Subsequently, 60% of LGBTQIA+ venues closed, and homophobic crime rates rose. People of lower income were then again forced out of the space they improved as developers and investors displaced the very people who helped improve the culture of the community.

Better design approaches for improving inclusivity and safety of marginalized communities have been done. One such case is the UK's public design space for people with different abilities. Three main points from these practices can be applied to improve safety for marginalized communities.

The first is to prioritize public safety. Practices that heterosexual couples take for granted, such as holding hands, can draw negative attention for people who are queer. Planners and designers can use green design and considerate lighting to break up public spaces without the use of harsh surveillance lighting. Vienna, Austria, pioneered gender-sensitive lighting to make areas feel safer and more comfortable to be inclusive of more people. This includes warmer lighting to encourage foot traffic and create semi-enclosed pockets in a park that offers some privacy but yet is still very walkable. With this lighting, people are more comfortable walking, thus reducing hate crimes.

Second, more effective inclusive design in areas needs to accommodate the specific needs of the LGBTQIA+ community. In addition to preserving venues and landmarks of significance to a culture, housing can be improved. Housing is typically designed for heterosexual families. Inclusive designs must account for all kinds of families and individuals in order to make communities more accessible. This includes creating communities for aging singles and those with health needs to have a support network.

Finally, making spaces visibly inclusive in design includes representation of queer heritage. This may include statues, memorials, and re-naming of streets, buildings and parks, public art, and other design approaches that celebrate a long rich past. Making visible design approaches such as artistic lighting, decorative street furniture, and queer plant communities may reduce hostility that people who are marginalized experience frequently.

These three approaches to design simply improve public spaces to make them safer and inclusive for communities of people who are marginalized.

6.4 Urban Streetscape with Public Space and Green Design

Expanding to the scale of a city block or neighborhood allows for more green interventions for equitable and inclusive design. Starting from a city street to understand design options and then moving to a larger scale of a block or neighborhood scale is beneficial for understanding how to see more

Image 6.12 **Urban areas of mixed use are opportunities for celebrating shared public spaces as areas of public assembly and encounters. Creating equitable and inclusive gatherings is an asset to the community.**

opportunities for equitable and inclusive design on a step-by-step basis. This design in Image 6.13 expands on the previous site design (Image 6.7) to increases the scale of the design interventions to include greening design on awnings, bus stops, balconies, and roof tops, as well as expanded public transportation, including buses with solar panels and living planted interior and exterior designs.

Pedestrian-scale lighting that is warm in color and consistent in spacing creates well-lit spaces for pedestrians to feel safe throughout the night. Design features that play a role in safety should be determined based on the needs of people who currently use the space, as well as those who do not use the space but wish to. People who do not use the space, but wish to, should be asked why they are not using the space and what can be done in design to encourage their use.

Sidewalks have zones for ease of use. There is a zone for features such as tree wells, which are expanded for further plantings and seating, trash disposal, streetlights, postal dropoffs, bus stops, bicycle parking structures, street names, community signage to celebrate events and people, public art, and other similar features. This same zone also has a curb cut for safe crossing of the street with colorful rumble strips. Beyond the sidewalk, there is a protected bicycle lane for cyclists and scooters, and then vehicle lanes.

Safe crosswalks of colorful street paintings and public art and signage celebrate people and communities that have been and continue to be underrepresented and underserved. Bringing these underrepresented and underserved voices to public spaces encourages the community to understand some of the effects toward people whose voices and needs have been – and are still – doubted, questioned, and treated as less than. As stated in earlier chapters, further actions beyond design – such as policy change, actions, support, education, and more – are also required to have lasting and effective change.

Urban neighborhoods include mixed-use spaces and buildings of mixed business and residential use. With people renting and owning their spaces in these communities, they become invested in the short-term and long-term health of their surroundings. Equitable and inclusive design created by people who are invested in their neighborhood is an asset for success.

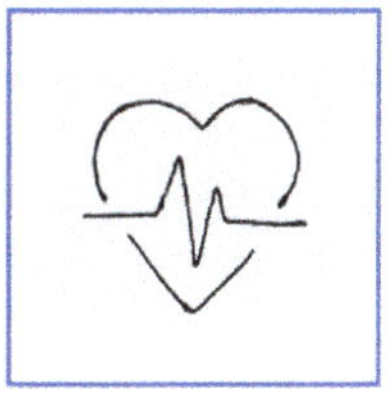

Health – Cycling lanes are protected with planted barriers for the safety of cyclists. Bikes are available throughout the community at no cost. The addition of expanded tree wells, planted awnings, rooftop greening, and creative green designs improves air quality for easier breathing. The addition of green design improves people's emotional health. Pedestrian safety is improved with wider sidewalks as well as a protective planted edge from vehicles.

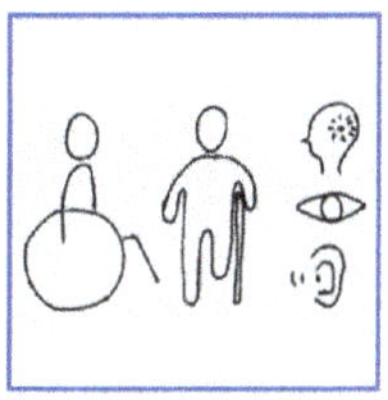

Abilities – Wide sidewalks for pedestrians, curb cuts, detectable rumble strips and warning pavers at corners, crosswalk melodies for walking and do not walk, and high-contrast signage improve navigation for people with different visual and walking abilities. A variety of seating options and spaces for people with different mobility skills are available.

Sexuality – Consistent luminescent lighting at pedestrian scale along each block, as well as warm lighting, improves safety for the LGBTQIA+ community. Signage of local businesses welcomes and celebrates people of this community.

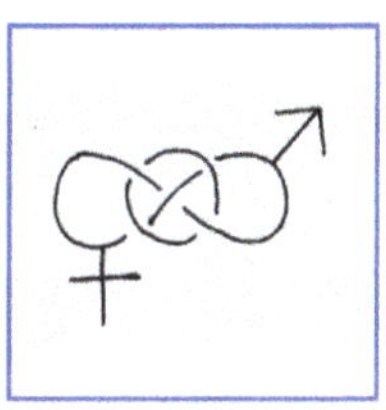

Gender – Public transportation is routed for womxn and people who run errands and take routes with multiple stops. Consistent luminescent lighting at pedestrian scale along each block, as well as warm lighting, improves safety for womxn in the community. Signage of local businesses welcomes and celebrates people of this community. Caregivers of children and senior adults have wider space for strollers and holding children's hands. Business signage, designs, street names, business merchandise, and services welcome and celebrate womxn.

Age – Curb cuts and safe crosswalks are at every street corner for people to safely cross streets at multiple locations on their route. This is important for people who walk throughout their day while running errands with children and people who are seniors. Fun designs are placed at lower heights for the sightline and tactile experience of children.

Race – Business signage, designs, street names, business merchandise, and services welcome and celebrate people who are Black, Brown, and from underserved communities.

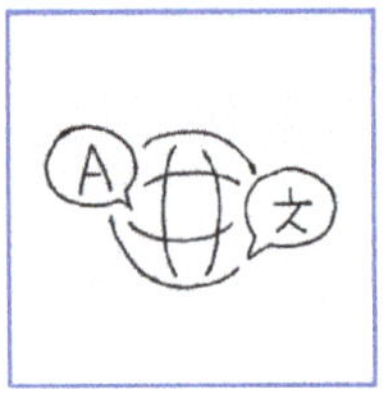

Language – Signs that use multiple local, regional, and international languages, as well as icons and images without words for those learning language, are used throughout the street at pedestrian heights.

Income – Public transportation is free or low cost for users who need it. Food-producing plants, which are designed throughout the streetscape, are available to eat to support people with low income in the community who want or need healthy food. The food can also be distributed locally to people or foodbanks.

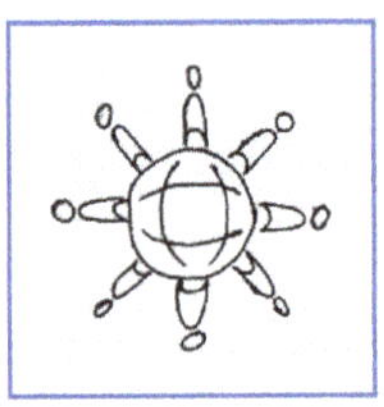

Culture – Business signage, designs, street names, business merchandise, and services welcome and celebrate people of different local, regional, national, and international communities.

Adding wide sidewalks, protected bike lanes, and green design creates public spaces of improved health and comfort for the community.

Image 6.13 Green design for a public urban streetscape.

An equitable and inclusive urban streetscape is layered with the needs and wants of large numbers of users. Creating shared spaces for pedestrians, cyclists, vehicles, buses, and animals requires planning ahead, as well as leaving space for change as materials, technology, and community needs change.

Image 6.14 **Poor design of a walkway that does not allow ease of walking up with a stroller.**

6.5 If Womxn Built Cities, What Would Our Urban Landscape Look Like?

Susanna Rustin explores the urban landscape through the eyes of gender, realizing that designers have considered the heights, perspectives, views, and wants of men (Rustin, 2014). One notable example of the urban landscape design is through the design of Le Corbusier. He explores working at the scale of the people. However, the challenge is that he worked with the height of six-foot-tall people. Generally, womxn and children are notably shorter than six feet.

Creating family-friendly and diverse spaces includes creating areas with ramps, more womxn's and family rest rooms, designing for the safety of

Images 6.15–6.16 **Image 6.15 is a public area with poor lighting, which creates unsafe walking routes in urban areas. Image 6.16 is designed with lighting that improves safety when walking.**

Images 6.17–6.18 **Creating green walkable cities with colorful living designs throughout the community creates healthy shared spaces. This style of design is both physically healthy for improving air quality and emotionally healthy with exposure to green design.**

womxn and people who are pregnant, and advocating for groups of people who need it. Beyond gender, consider a diverse range of people based on ethnicity, socioeconomic background, and age, and other people who are marginalized – such as teenagers, for example.

There are different experiences in an urban landscape that is designed for six-foot-tall men, when one is smaller, pregnant, feeling unsafe after dark, or feeling marginalized due to age, abilities, or on some other basis. Rustin writes that urban designs would be different if more voices were heard and involved in the process.

In addressing gender, strategic practices such as political representation and recognition of the spaces are not just about buildings and open space, but importantly are about how people are able to be and feel within the space that supports the needs and wants throughout the lifetime of womxn.

As a note from this author: I have seen design practices and green spaces change throughout my time in design. Though there have been some changes

Images 6.19–6.20 **Image 6.19 illustrates the lack of comfortable seating that suits womxn who are breastfeeding. Image 6.20 is a lactation station for comfortable seating when breastfeeding.**

toward inclusive design and growth in representation of underrepresented and marginalized groups in design fields, *better is never enough*. Slow change serves to acknowledge that change is needed, known to be needed, yet the practice of "designed by and for men" is still in high favor. *Equitable and inclusive are always the options to choose*. Small changes over time prove that change can happen – and if change can happen, let us go for full equity and inclusion, rather than frustratingly see "better but still not inclusive" practices and designs. Let us choose full equity and inclusion, immediately and moving forward.

6.6 Multiple Lane Road with Public Space, Safety, and Green Design

Creating a space for pedestrians and vehicles moving together requires weaving safety and creativity throughout layers of the design. The green design in Image 6.22 considers pedestrians, drivers, cyclists, community, public comforts, and healthy spaces together to retrofit with equity and inclusion as its goals for safety.

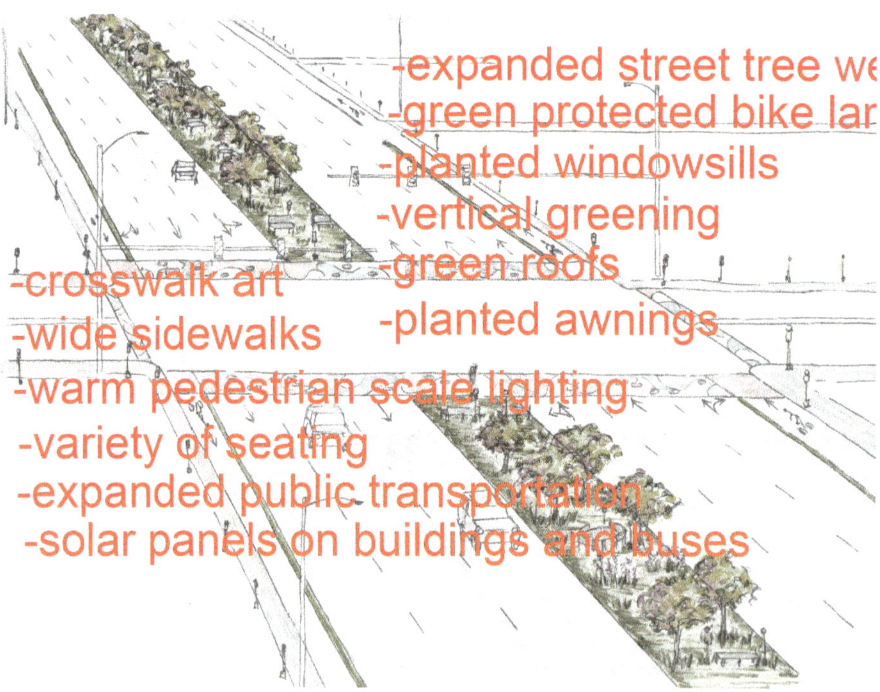

Image 6.21 Multiple-lane roadways with safe healthy green design help weave pedestrian-scale spaces with a larger scale. Combining different scales of spaces, different modes and speeds of transportation, and a variety of users is required for equity and inclusion.

In a highway design, integrating pedestrians, cyclists, drivers, and greening creates a weaving of multiple components toward equity and inclusion for public comfort and safety. This design has wide sidewalks with pedestrian-scale lighting, crosswalk buttons with sounds for people with different vision abilities, green-covered bus stops, algae biopanels, and expanded tree well gardens. The protected bicycle lanes have a planted green edge for safety, as well as improving air quality.

The colorful crosswalk is painted with air-purifying paints and murals that celebrate underrepresented and underserved communities. Revealing their stories in public art creates awareness of some issues of inequity. Awareness is one role that is necessary for change.

The median with greening design serves multiple purposes. First, it serves as a stopping point for pedestrians who need the space to stop in the crosswalk. Pedestrian-scale lighting in the median is for use at night. There is seating of multiple styles for a variety of users. For safety of the vehicles, the green median blocks some vehicle headlights, which makes driving safer for drivers on the opposite side of the roadway. The green median is slightly raised above street level for pedestrian safety.

Very notable in this design are a tree orchard, shrubs, and plants that produce food. Providing food and public spaces allows people who need them to have access to them. Public access to healthy food supports inclusion for community members to be caretakers of the space, as well as people to distribute the food to those who want or need it.

Weaving green design throughout this site for equity and inclusion requires safety, creativity, and consideration of needs of diverse users. Improving people's emotional and physical health supports long-term community success.

Health – Street murals use air-purifying paint, which absorbs and neutralizes chemicals, pollutants, and VOCs for improved air quality. The center median has a lush green design to improve air quality and create a space for stopping, sitting, or community gathering. The center area may be raised or at street height for safety of the pedestrians – as well as the drivers – to reduce seeing headlights from vehicles on the opposite side at night. Cycling lanes are protected with planted barriers for the safety of cyclists.

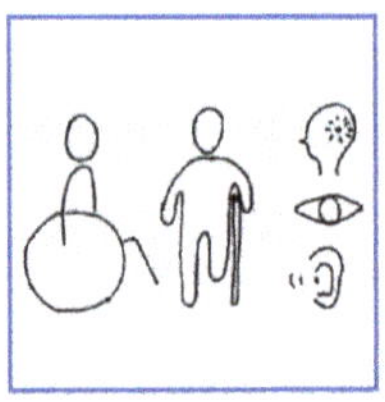

Abilities – The center median provides a safe resting space for people who need a longer time for crossing. The street murals are of colors visible to people with different visual abilities. Curb cuts, detectable rumble strips and warning pavers at corners, crosswalk melodies for walking and do not walk, and high-contrast signage improve navigation for people with different visual and walking abilities. A variety of seating options and spaces for people with different mobility skills are available.

Sexuality – Consistent luminescent lighting at pedestrian scale along each block, as well as warm lighting, improves safety for the LGBTQIA+ community. Signage of local businesses welcomes and celebrates people of this community.

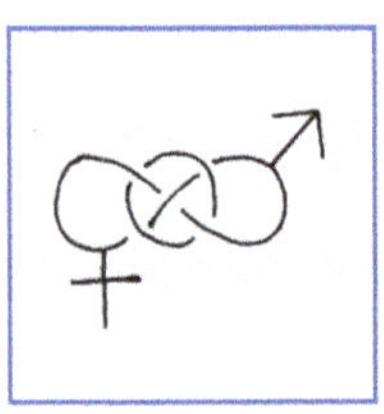

Gender – The center median allows a resting or stopping space for caregivers with children and senior adults. Consistent luminescent lighting at pedestrian scale along each block, as well as warm lighting, improves safety for womxn in the community. Public transportation is routed for womxn and people who run errands and take routes with multiple stops. Designs, signs, and street names celebrate womxn.

Age – The center median allows a resting or stopping space for people who need time to cross the street. Curb cuts and safe crosswalks are at every street corner for people to safely cross streets at multiple locations in their route. This is important for people who walk throughout their day while running errands with children and people who are mature.

Race – Communities of people who are Black, Brown, and underserved, which have been disconnected by roadways, are connected to their surrounding communities with a safe pedestrian green belt. Designs, signs, and street names celebrate people who are Black, Brown, and from underserved communities.

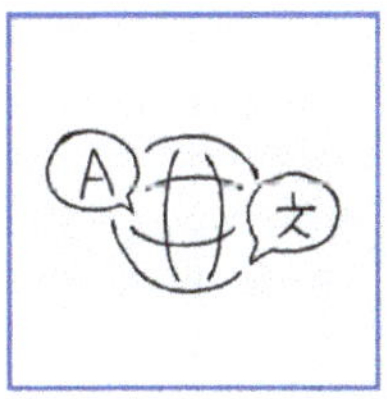

Language – Signs that use multiple local, regional, and international languages, as well as icons and images without words for those learning language, are used throughout the street at pedestrian heights.

Income – Communities of people who have low income, which have been disconnected by roadways, are connected to their surrounding communities with a safe pedestrian green belt. The green median has food-producing plants that are available to eat to support people with low income in the community who want or need healthy food. The care for the median can also provide jobs to people who want temporary or long-term jobs.

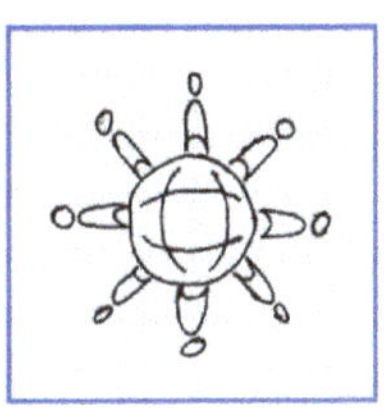

Culture – Communities of people of different cultures, which have been disconnected by roadways, are connected to their surrounding communities with a safe pedestrian green belt. Designs, signs, and street names celebrate people of different local, regional, national, and international communities.

Creating an urban area that is safe during the daytime and night-time for equity and inclusion further creates a stronger community.

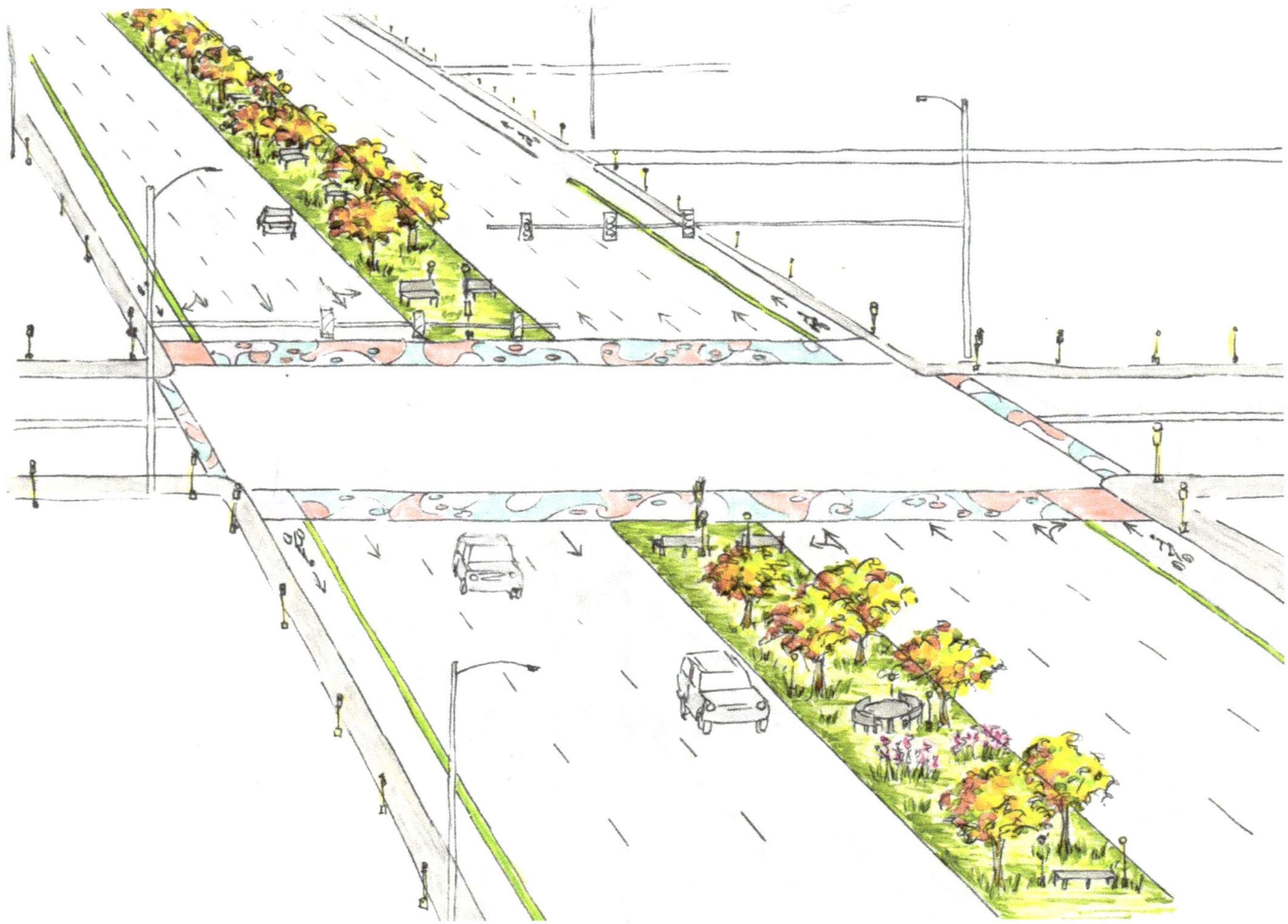

Image 6.22 **Healthy green streetscape designed for inclusion of people and vehicles.**

An equitable and inclusive streetscape is layered with the needs and wants of a large number of users. The layers of shared spaces for pedestrians, cyclists, vehicles, buses, and animals requires planning ahead, as well as leaving space for change as materials, technology, and community needs change. The safety of all of the users as well as an enjoyable experience is required for equity and inclusion.

Image 6.23 **Womxn are 55% of the users of public transportation in the United States.**

6.7 More Womxn Ride Mass Transit Than Men – Shouldn't Transit Agencies Be Catering to Them?

A 2007 study from the Public Transportation Association found that 55% of the overall ridership of public transportation across the United States is womxn (Goodyear, 2015). In some parts of the country, this percentage is even higher. A survey by Philadelphia's city transit agency, SEPTA, found that

Image 6.24 **A 2007 study found that 64% of the users of Philadelphia's public transit are womxn (Goodyear, 2015). Creating inclusive public design equals truly addressing the public users – in this case, womxn.**

Image 6.25 **Redesigning public transit for womxn includes lighting for safety and schedules for daytime and night-time routes when people are running multiple errands – often with children or people they are caring for.**

64% of the users of subways and buses are womxn; Chicago's survey found that 62% of the riders are womxn; and Washington, DC; New York; and Boston state that 60% of their users are womxn.

However, the survey also found that the routes and schedules of the transit do not cater to the womxn, the majority of their riders. Further, researchers such as Anastasia Loukaitou-Sideris, a professor of urban planning at UCLA,

Image 6.26 **Addressing the primary users of public transit, who are womxn and children, requires routes, seating, safety, and schedules suitable for these primary users.**

have found that womxn have consistently said that they are concerned for their own safety on public transit (Loukaitou-Sideris, 2009).

In the United States, womxn's concern for safety is rarely addressed by administrative and design solutions in public transportation, leaving womxn to find their own solutions and cope with it on their own. This often leads to these womxn taking different travel patterns, which extend their routes and time of travel. Further, with crosswalks not available at most bus stops, womxn who have children in tow risk greater safety issues, either crossing without a crosswalk with traffic or walking further distances out of the way just to get to a crosswalk while having strollers, children, or people with wheelchairs in tow.

European countries, as well as Australia, Canada, and Japan, have begun to address this inequality of public transit by launching programs specifically targeted at womxn's safety, as expressed by womxn. Further, womxn are presently the predominant caregivers of children, as well as being responsible for most or all of the household errands. Improving accommodations for strollers, wheelchairs, and ease of travel with multiple people of varying ages is a design and practice that improves the lives of womxn – the primary users of public transportation.

6.8 Urban Community Space Under Highway, Green Design

Communities of people who live or work near freeways, highways, and train routes are greatly affected by them. The site in Image 6.28 utilizes design

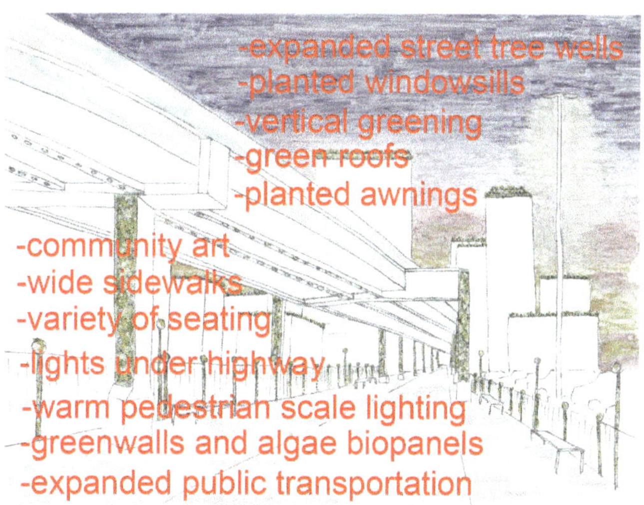

Image 6.27 This design creates space for communities to walk, gather, and celebrate under highways that once divided them. Removing barriers to create healthy green spaces for people returns connections and sharing of neighborhoods that have been intentionally separated.

interventions for safe, welcoming, celebrated inclusive space that removes vehicular traffic under the highway in order for people within the community to walk and reconnect with neighbors who have been separated by a major roadway.

On the ground, as well as on other surfaces, murals are painted with air-purifying paint for improving air quality. Public murals with air-purifying paint can be applied to the roadway columns and a variety of other locations throughout the site and can change as the community does. A variety of seating options for people of different heights and abilities are used throughout the space, as well as other basic comforts such as trash cans and the use of surfaces that are easily walkable for people with strollers or walkers, or in wheelchairs; people with different walking abilities; and people who wear shoes with a variety of soles or fashions.

Signage is integrated throughout the site, using words in multiple languages, as well as symbols to explain the intended message.

In order to reduce the effects of the loud sounds of traffic, plants are used to reduce sound waves. This includes expanded tree well gardens, as well as green roofs, vertical green walls, and algae biopanels. Greening is designed into public art pieces, fence posts, and highway columns to improve air quality and be utilized for energy. The energy can support the lighting under the highway. Lighting under the highway is also supported by solar panels added to streetlights. The design of the lighting is warm and colorful, and can change based on events and celebrations. In addition to the overhead lights are pedestrian-scale lights throughout the site.

Creating sites that prioritize pedestrians in urban areas encourages people to take time to know their community at a walking pace. These spaces also expand where people of mixed health, abilities, sexuality, gender, age, race, language, incomes, and cultures can be together and grow, heal, learn, laugh, and live with more joy and health.

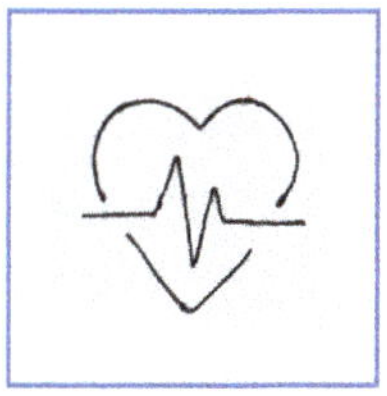

Health – Lighting under the highway creates a safe space for night use for pedestrians and cyclists with the area closed for vehicular use. Algae panels on the vertical surface, rooftop greening, and creative green designs improve air quality for easier breathing as well as emotional health with planted areas in urban areas at pedestrian scale. Street murals use air-purifying paint, which absorbs and neutralizes chemicals, pollutants, and VOCs for improved air quality.

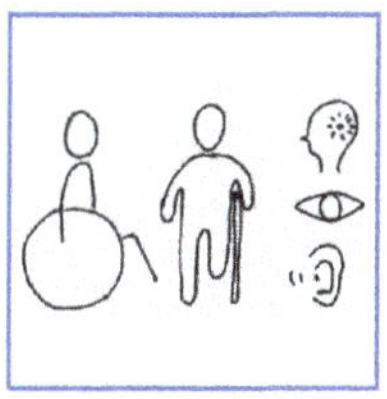

Abilities – Streets for pedestrian use (no vehicles), wide sidewalks for pedestrians, curb cuts, detectable rumble strips and warning pavers at corners, crosswalk melodies for walking and do not walk, and high-contrast signage improve navigation for people with different visual and walking abilities. A variety of seating options and spaces for people with different mobility skills are available.

Sexuality – Consistent luminescent lighting at pedestrian scale along each block, as well as warm lighting, improves safety for the LGBTQIA+ community. Designs, signs, and street names celebrate people of this community.

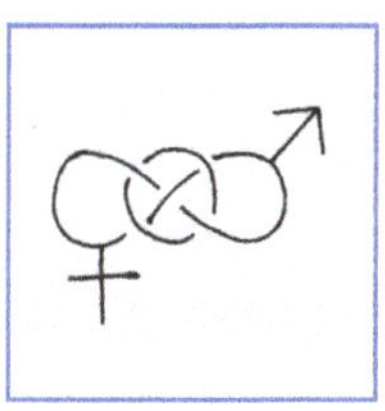

Gender – Consistent luminescent lighting at pedestrian scale along each block, as well as warm lighting, improves safety for womxn in the community. Signage of local businesses welcomes and celebrates womxn of this community. Caregivers of children and senior adults have wider space for strollers and holding children's hands. Designs, signs, and street names celebrate womxn.

Age – Seating for a variety of people with different heights and mobility are throughout the space. Fun designs are placed at lower heights for the sightline and tactile experience of children.

Race – Communities of people who are Black, Brown, and underserved, which have been disconnected by roadways, are connected to their surrounding communities with a safe pedestrian green belt. Designs, signs, and street names celebrate people who are Black, Brown, and from underserved communities.

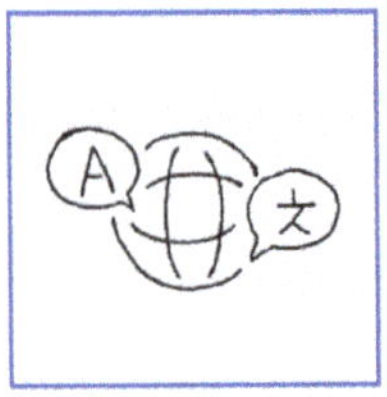

Language – Signs that use multiple local, regional, and international languages, as well as icons and images without words, for those learning language, are throughout the street at pedestrian heights.

Income – Communities of people with low income, which have been disconnected by roadways, are connected to their surrounding communities with a safe pedestrian green belt. Designs, signs, and street names celebrate people of different local, regional, national, and international communities.

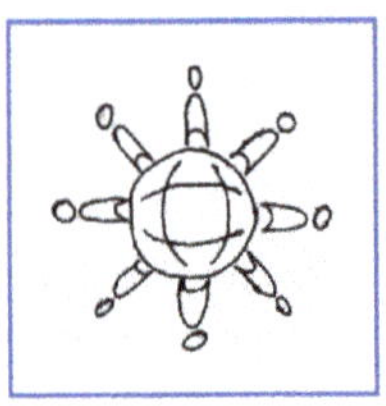

Culture – Communities of people of different cultures, which have been disconnected by roadways, are connected to their surrounding communities with a safe pedestrian green belt. Designs, signs, and street names celebrate people of different local, regional, national, and international communities.

Creating an urban area that is safe during the daytime and night-time for equity and inclusion further creates a stronger community.

Image 6.28 **Green space for pedestrians under freeways and trains.**

Creating green public spaces in urban areas require unique approaches to underutilized areas. Adding a green pedestrian path at this site includes using algae biopanels and views to green roofs for improved physical and emotional health. The green site is shared by a diverse community for both safe and healthy daytime and night-time use in a dense urban area.

Image 6.29 **Redesigning staircases with ramps integrated into the design for pushing strollers, or for use by people who use wheelchairs or walkers.**

6.9 City With a Female Face: How Modern Vienna Was Shaped by Womxn

The Aspern neighborhood of Vienna, Austria, is expecting its family-oriented redesign to be complete by 2028, though presently it already has one of the highest qualities of living of all the cities of the world (Hunt, 2019). Gender mainstreaming is part of the inclusive redesign of Aspern. Redesigning Aspern with a focus on the needs of womxn started more than 30 years ago through gender mainstreaming. Gender mainstreaming is the practice that ensures womxn are accounted for equally in policy, legislation, and resource allocation. This policy was proposed in 1985 and has taken decades to be considered.

Image 6.30 **Vienna, Austria, street setting with walkable streets, street trees, bike racks, public comforts, lighting, and spaces with community events.**

Images 6.31–6.32 **Warm safe lighting creates walkable spaces for night use in Image 6.31. Image 6.32 is a walkable street at pedestrian scale with mixed-use spaces.**

The practice of gender mainstreaming is that you get a fairer society and better quality of life for more people. Two notable foci are transportation and safety. One example of a city design practice is that male plans focus on the system that there would be no other people than men going to work in the morning and coming back in the evening, and everything else in between – such as childcare, unpaid labor, shopping, errands, and short trips on foot or by public transportation – is ignored.

A study by the Social Democratic Party of Austria (Solsten and McClave, 1994) revealed that two-thirds of the automobile trips are by men, and two-thirds of the foot trips are by womxn. This leads to a very different city experience, which was being ignored in city design. Womxn feel unsafe in the city design, unable to easily get around, and not included in the spaces.

A program in 2002–2006 in Mariahilf, Vienna, created 26 improved spaces with gender mainstreaming. The area altered traffic light patterns to prioritize pedestrians, added public seating, widened walkways, and removed barriers

Images 6.33–6.34 **The Vienna, Austria, street shown in Image 6.33 is named Janis-Joplin-Promenade (named after a notable womxn). Image 6.34 illustrates the plazas, parks, and walkable areas in Vienna, which create spaces designed by womxn and for womxn, creating designs which are welcoming, safe, and intentionally inclusive.**

Images 6.35–6.37 These are not the original case study photos, but rather are recreated images to represent how photos were used in the Gender-Based Campus Walking Study.

Using a heat mapping tool to note the areas that caught their attention – orange circles represent areas where people spent looking at most of their time. Blue circles represent little time and focus spent understanding the space.

The first image is a site at night with a central walking path. Image 6.36 illustrates how womxn viewed the space. Womxn intensely understand the space by seeing around the entire edge of the space, including the poorly lit dark areas – as illustrated by the dark orange circles. Notably less time and attention was spent on the central path as the primary focus, as noted by the blue circles.

Image 6.37 illustrates how men viewed the space. Men intensely focused on the central direct path, as noted by the orange circles, and spent very little time on the peripheral scene, as noted by the blue circles.

These images illustrate that there is a strong difference between how womxn and men understand and experience spaces for their safety.

for people in wheelchairs or walkers. City infrastructure and services are best served when they are inclusive. Gender budgeting, introduced in 2005, requires each division to report how their expenditures benefit womxn and men equally, with sanctions for those who do not comply. The goals of these practices are to make womxn's lives more equal.

Gender mainstreaming was – and still is – met with resistance from both lower and higher levels of administration, as well as the public. If gender mainstreaming is not applied, those of the dominant group would be most invited to use the spaces, and others would not feel welcome, continuing a decline in minority groups and continuing the inequitable cycle. When practiced well, gender mainstreaming comes across as simply a well-designed neighborhood rather than as a feminist utopia.

In Aspern, Vienna, all the streets and public spaces are named for notable international womxn, which is counter to traditional Vienna, where 3,750 streets are named after men. This was met with resistance by some. It is interesting that when 3,750 streets were named for men, there was not the same resistance. But when forced to see inequity in smaller and larger city

Images 6.38–6.39 Womxn tend to avoid walking near dense barriers; they prefer open pathways, warm well-lit areas, or areas where there is clear visibility to the surrounding spaces for their safety.

practices, resistance is voiced by those who fear their comfort will be lost in order for others' comfort to be gained.

The day must be today when "more equal" is not ever a part of the equation. Equitable is the only standard that is acceptable – ever.

6.10 Gender-Based Heat Map Images of Campus Walking Settings: A Reflection of Lived Experience

In a Brigham Young University study published in the journal *Violence and Gender*, Chaney and colleagues (Chaney et al., 2024) address how womxn understand and experience spaces they walk in, as well as how that is different from how men experience spaces. Using students who are womxn and men, the 571 participants (276 womxn, 219 men) of the study were asked to look at 16 different images of areas on the campus. They were told to imagine themselves walking through the space. The participants were asked to click on areas of the photos that caught their attention.

The heat map images illustrate the vast differences between how men focus on the path directly ahead while womxn notably focus on peripheral spaces, potential hazards, the surrounding areas, and dark areas and objects on and off the path.

> Our view of and experience with the world can have far-reaching influence on our health, both in the short and long run. For example, fear can decrease motivation for physical activity (Foster et al., 2014a; Stodolska et al., 2013). Multiply this over the course of one's college experience, or adult life, and the results are troubling (Poobalan et al., 2012; Telama et al., 2005). At times, fear unfairly burdens those it afflicts by limiting their power to pursue life, liberty, and happiness (Kennedy, 2023; Rierson, 1994).

> Fear associated with night-time, or poorly lit, commuting has been shown to negatively affect the physical, emotional, and social health of female students (Burdette and Needham, 2012; Larson et al., 2013; Park and Garcia, 2020). In addition, research shows safety perceptions can greatly affect sleep patterns and academic performance (Etopio et al., 2019).
>
> Bolstering community walking conditions around college campuses is imperative, owing to the link between freedom to move around a community and basic health outcomes (Boomsma and Steg, 2014; Peachey and Baller, 2015). Gaining an understanding of the factors that affect the safety of female college students can give evidence to drive community improvements within the college community. By recognizing and assessing the influence of the built environment and safety perceptions on female college students' walking autonomy, we add information about needed environmental and systems changes within U.S. college towns to promote improved female college student health and well-being.

A note from this author: As a designer of spaces and a womxn, it continues to surprise me that the knowledge that womxn and men understand spaces differently is still not understood well, and further, that spaces have been primarily designed by men and for men based on height, lighting, use, scale, walkability, and safety – to name a few. Health and safety for womxn, people in the LGBTQIA+ community, and people with different abilities and ages are not prioritized. *When equitable and inclusive design is the standard, culture as a whole rises. When equitable and inclusive design is the standard, more users are acknowledged, respected, and welcomed into public and private spaces. Equitable and inclusive design is the standard for success – nothing less is acceptable for culture.*

6.11 Interior Agriculture and Park, Green Design

This indoor park in Image 6.41 inside an abandoned warehouse or industrial building creates a year-round space of fresh-growing greening with changing spaces for play and events. Utilizing abandoned structures for community design is a creative way to retrofit greening in neighborhoods where greening was not programmed in earlier stages. The space includes indoor sinks, a kitchen for teaching, and other community needs for enjoying the space, such as a variety of seating options, tables, games, books, artworks, and more.

In this space, there is a stage for music, theater, or other venues to perform. There is also a play area for children, and floor games such as bocce ball. The larger scope of the site is a park with trees, shrubs, green walls, grasses, flowering plants, and food-producing plants with lighting and irrigation to support the green design. Stormwater from the roof and greywater from the building can be utilized for irrigation of the park.

Caretaking the park includes professionals and is also a space for training and teaching people. With a variety of plantings, from food-producing plants to year-round greening, the opportunities to use the space as a teaching resource for the community is open. Distribution of the food produced in the site is also available for people who need it, and it can be shared with the community.

Retrofitting for social equity improves quality of life and health. Creating indoor parks supports public comforts for communities year-round and offers a direction for healthy green design of equity and inclusion. In areas where people live next to unused buildings and spaces, the option to create healthy green design addresses an approach to the built space that that CDC with the SDOH (see Chapter 1 for more information on the SDOH) supports for quality-of-life health. Creating built structures for health improves social equity and addresses health as a right for all people, animals, plants, and earthly spaces.

This interior park creates a healthy quality of life for users of the community to share in short-term and long-term growth toward success. In an interior green park, weather conditions do not interfere with coming together for events or daily walks.

Image 6.40 **Utilizing an empty warehouse or large space for an indoor park allows year-round use for a community. With a venue for events such as small plays or music, as well as green design, the community can gather, share, learn, and enjoy the space through warm and cold outdoor weather.**

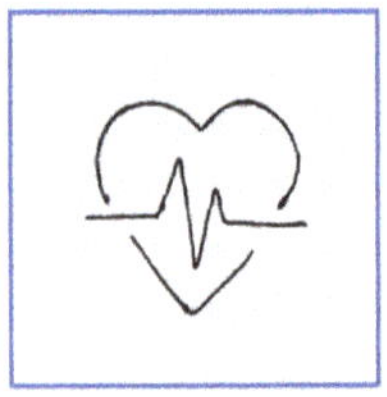

Health – The year-round green space allows people to enjoy healthy air quality, fresh plant scents, colorful views, a variety of textures to touch, and food to eat. The space improves emotional health with access to a public interior green area open year-round for unique growing and changing community spaces.

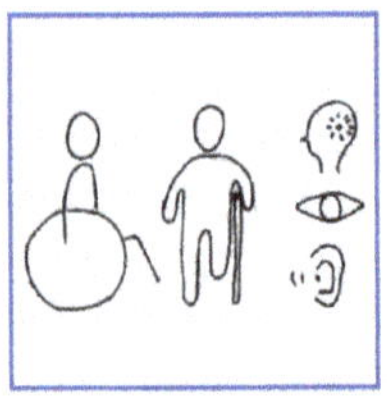

Abilities – Wide pathways, movable features, and high-visibility signage allow comfortable spaces for people with different walking and visual abilities. A variety of seating options and spaces for people with different mobility skills are available.

Sexuality – Consistent luminescent lighting at pedestrian scale is throughout the space, as well as warm lighting to improve safety for the LGBTQIA+ community. Design and signs welcome and celebrate people of this community.

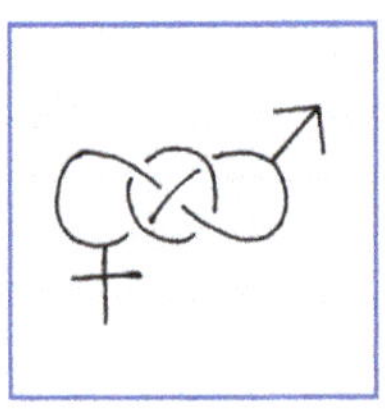

Gender – The mixed-use design allows spaces with views for caregivers to oversee children as they play in the space. Consistent luminescent lighting at pedestrian scale is throughout the space, as well as warm lighting to improve safety for womxn. Designs and signs welcome and celebrate womxn in this community. Caregivers of children and senior adults have wider space for strollers and holding children's hands.

Age – An interior green site creates a mixed-use space for children, teenagers, adults, and seniors to share. Design features are placed at a variety of heights with tactile and scent experiences for a variety of users. The shared space expands the community for a rich tapestry of users. With set structures as well as movable design components, the space continually changes for diverse uses.

Race – Using abandoned or underutilized spaces for healthy green design serves Black, Brown, and underserved communities, which have been disconnected from the community green spaces. Designs and signs welcome and celebrate people who are Black, Brown, and from underserved communities.

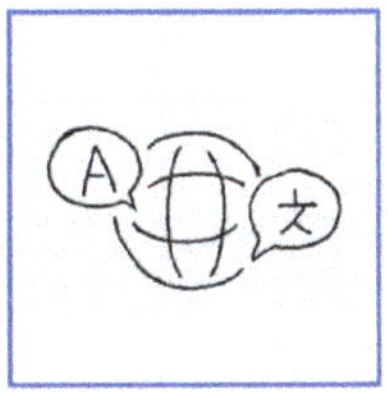

Language – Signs that use multiple local, regional, and international languages, as well as icons and images without words, for those learning language are throughout the street at pedestrian heights.

Income – Food-producing plants designed throughout the site are available to eat to support people with low income in the community who want or need healthy food. The food can also be distributed locally to people or foodbanks. The care for the interior green space can also provide jobs to people who want temporary or long-term jobs.

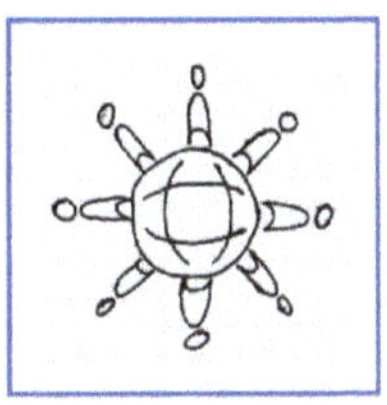

Culture – Communities of people of different cultures have a shared healthy green space to gather year-round. Plants and designs welcome and celebrate people from different local, regional, national, and international communities.

Indoor green public comforts encourage a community of diverse people to gather throughout a year in a healthy space.

Image 6.41 An interior park for long-term green design, as well as changing events.

Interior green parks and green venues have long-term and large-scale greening, as well as spaces for mobility with changing events. Making sure the green design allows both to work together benefits the community to have a space for both the community and visitors. The mix of people encourages visitors to bring this idea to their community after experiencing it.

Images 6.42–6.43 Creating clear signage with multiple languages, as well as symbols, allows a greater range of people to be able to navigate and use a space safely and independently.

6.12 What Three Weeks in Japan Taught Me About Public Transit Signage

In an article by J. Tauberer (2020), the design approaches that are successfully used in Japanese train stations can be used for improved inclusivity. Specifically, design with consideration of multiple language options and different abilities is considered.

The signage, both digital and stationary, uses multiple languages in large font. Names and numbers in large Arabic numerals (0, 1, 2, 3 . . .) and Roman numerals (I, V, X, L, C, D, M representing values of 1, 5, 10, 50, 100, 500, 1000, respectively) improve inclusivity – not only for language but also for people who have vision less than 20/20. In addition to words, large symbols are also used to express navigation through the extensive and busy train stations. The symbols are large and use color consistency. For example, yellow is consistently used for exits from the station throughout all the stations in the system. Inclusive design continues in these train stations with the use of specific colors for each of the train lines.

Images 6.44–6.45 Colors, size of text, and contrasting colors and shapes are design approaches that create inclusive designs for users to navigate spaces enjoyably.

Images 6.46–6.47 **Inclusive design can utilize technology with lights that turn on or change color as people move, as well as with quick-response (QR) codes for use with handheld devices.**

To make the spaces safer for the quick pace of a high density of people, colorful floor decals indicate where people are to stand or line up if they are getting on or off a train. In tight spaces, the floor decals make turns to indicate a safe path for a growing line of people. There are also decals of where not to stand for safety purposes. Both floor and wall decals show navigation for directions of walking up or down staircases.

In addition to colors and language, the sound of long musical jingles lets people prepare for the doors closing. The jingles are much longer in time than those on U.S. trains, which are very quick. Different trains also have different jingles, so one can hear if they are on the correct train by the jingle sound. For example, the train for Disney in Japan plays a jingle of "It's a Small World."

The use of language, numbers, symbols, font size, colors, and sounds creates an inclusive design in the public transit system of Japan. Many of these approaches can be applied to other public sites.

Images 6.48–6.49 **High-contrast symbols on walkways are one approach for people of different visual abilities. Use of symbols on the walkways, rather than text, is inclusive for people who speak different languages to be able to navigate the space.**

Images 6.50–6.51 Rescaling structures for shared use of people who are mature, as well people who are young, encourages shared spaces for community health.

6.13 Playground for Seniors Coming Your Way and Ballum Bumps Playground

New playgrounds can be designed for people of a greater range of ages. There is a style of playgrounds called "free playgrounds" designed for seniors (Abbit, 2015). Most playgrounds that are designed for seniors feature low-impact exercise apparatuses, as well as features designed for flexibility, balance, coordination, and encouraging community gatherings. Most importantly, they are designed to be fun!

Free playgrounds for mature people have a few different names, including outdoor fitness parks and senior playgrounds. These were designed in the mid-1990s in China as they required fitness programs for people of all ages. By the 2000s, these fitness areas spread to England, Japan, Finland, Germany, Canada, and Spain, as more that 40% of the population will be 65+ by 2050. As of 2023, Barcelona has more than 300 senior playgrounds. Another example of an unstructured play space for toddlers through seniors is Ballam Bumps Regional Play Space, located in Karingal, Australia (AILA, 2018). Senior playgrounds as well as combined senior and children playgrounds can improve people's physical health and emotional well-being.

Images 6.52–6.53 Creating outdoor spaces for physical health of people who are mature is an inclusive design approach for physical and psychological health.

Images 6.54–6.55 **Inclusive areas for teenagers and people who are mature are a relatively new practice in design, as teenagers are a typically overlooked population.**

The United States has started to build senior playgrounds and multi-generational fitness parks. Michael Cohen is the founder of Must Have Play, the first U.S. company dedicated to wellness playgrounds for seniors. These designs include stationary and recumbent bikes; side-by-side striders; face-to-face leg presses; hand-eye dexterity games; walking paths with ramps, steps and arches; bocce ball; ping pong; and horseshoes.

As a green design specialist, I have found that the teenage population has been generally overlooked in design. Beyond the needs of children, and not as independent as adults, this stage of growth is underrepresented in spatial design. Intentional design, which has the interest and scale of teenagers woven into spaces with people who are younger and older, helps teens to find their voice and interests, and helps them grow in healthy ways.

Images 6.56–6.57 **Playground structures that are inclusive for children in wheelchairs are mixed with structures for children who are not in wheelchairs, rather than having separate structures for different users.**

Images 6.58–6.59 **Images of a braille sign in a building placed in an unsuitably high location, making it out of sight for people with different visual abilities to logically know where to reach to find it. The womxn in the photograph is 5-foot-7, with the braille sign out of reach.**

6.14 Moving Forward With Public Comforts and Safety

Effective and enjoyable design is the difference between basic compliance and thoughtful inclusive spaces. Real-world sites create unique opportunities for applying green designs for people toward their health, abilities, sexuality, gender, age, race, language, income, and culture. Further, people understand and experience spaces differently based on these. Yet, design has not caught up to be inclusive of these differences, prioritizing and favoring healthy cisgender Caucasian men. This results in spaces that are not equitable for others.

Through my life, I have found that womxn are taught to understand and look at spaces differently for their safety. Meanwhile, cisgender men are not taught this. People are taught different standards based on gender – and that is a larger challenge that culture has to change. In addition to making spaces safe and inclusive of the needs of womxn, there is also a need to teach culture about equitable standards of safety.

For example, green design cannot address the fact that in the United States, 91% of victims of rape and sexual assault are womxn, and 9% are men; with men as 99% of the perpetrators of the sexual assaults and rapes (North Coast Rape Crisis Team, n.d.). Teaching the larger society to treat people equitably and with a mindset of inclusion is needed to address this.

This does not stop with gender standards. Similarly, yet differently, there are ever-changing and varying standards which are applied to race, culture, health, abilities, sexuality, income, age, and language.

Equitable and inclusive design cannot prioritize a one-size-fits-all solution over true inclusivity. True inclusivity requires designs that consider the wants and needs of all users and encourage all users to use the space independently and comfortably. While green design changes to address spaces, culture must also continue to change to address the inequitable and exclusive standards that are far too deeply embedded and practiced in a culture. If equity and inclusion make someone uncomfortable, the issue of their own fears is within themselves and cannot determine equity and inclusion going forward. The health of culture requires the strength of people to create immediate and long-term solutions, in both green design and beyond. This book gives you, the reader, applications to be a part of that strength. Use it with all the wisdom, creativity, and joy that you can – and go forward!

Image 6.60 Creating spaces for communities to gather year-round in all neighborhoods, particularly those that have been underserved, serves an important role for the quality of life for people who live and work in healthy, safe, and enjoyable surroundings.

References

Abbit, L. (2015). *Playgrounds for Seniors Coming Your Way*. Senior Planet, AARP. Retrieved 20 December 2023, from https://seniorplanet.org/playgrounds-for-seniors/

Australian Institute of Landscape Architects [AILA]. (2018). Ballum Bumps Regional Play Space. *AILA 2019 National Landscape Architecture Award*. Retrieved 20 December 2023, from https://www.landscapearchitectureprojects.com/projects/2018/ballumbumps

Boomsma, C., & Steg, L. (2014). Feeling Safe in the Dark: Examining the Effect of Entrapment, Lighting Levels, and Gender on Feelings of Safety and Lighting Policy Acceptability. *Environmental and Behaviour*, 46(2), 193–212.

Burdette, A. M., & Needham, B. L. (2012). Neighborhood Environment and Body Mass Index Trajectories From Adolescence to Adulthood. *Journal of Adolescent Health*, 50(1), 30–37.

Cattarall, P., & Azzouz, A. (2021). The Queer City: How to Design More Inclusive Public Space. *The Conversation US*. Retrieved 20 December 2023, from https://theconversation.com/the-queer-city-how-to-design-more-inclusive-public-space-161088

Chaney, R., Baer, A., & Tovar, A. L. (2024). Gender-Based Heat Map Images of Campus Walking Settings: A Reflection of Lived Experience. *Violence and Gender*, 11(1), 35–42. Retrieved 1 June 2024, from https://www.liebertpub.com/action/showCitFormats?doi=10.1089%2Fvio.2023.0027

Chapman, R. (n.d.). *Parks and an Equitable Recovery*. Trust for Public Land. Retrieved 16 February 2025, from https://www.tpl.org/parks-and-an-equitable-recovery-parkscore-report#:~:text=Our%20data%20shows%20major%20disparities,in%20park%20investment%20and%20planning

Etopio, A. L., Devereux, P., & Crowder, M. (2019). Perceived Campus Safety as a Mediator of the Link Between Gender and Mental Health in a National U.S. College Sample. *Women Health*, 59(7), 703–717.

Foster, S., Giles-Corti, B., & Knuiman, M. (2014a). Does Fear of Crime Discourage Walkers? A Social-Ecological Exploration of Fear as a Deterrent to Walking. *Environmental and Behaviour*, 46(6), 698–717.

Goodyear, S. (2015). More Women Ride Mass Transit Than Men. Shouldn't Transit Agencies Be Catering to Them?. *Bloomberg*. Retrieved 20 December 2023, from https://www.bloomberg.com/news/articles/2015-01-30/more-women-ride-mass-transit-than-men-shouldn-t-transit-agencies-be-catering-to-them

Hunt, E. (2019). City with a Female Face: How Modern Vienna was Shaped by Women. *The Guardian*. Retrieved 20 December 2023, from https://www.theguardian.com/cities/2019/may/14/city-with-a-female-face-how-modern-vienna-was-shaped-by-women

Kennedy, J. (2023). Power to Pursue Happiness. *Daedalus*, 152(1), 52–57.

Larson NI, Wall MM, Story MT, et al. (2013). Home/family, peer, school, and neighborhood correlates of obesity in adolescents. *Obesity*. 21(9), 1858–1869.

Loukaitou-Sideris, A. (2009). *How to Ease Women's Fear of Transportation Environments: Case Studies and Best Practices*. Retrieved 25 February 2025, from https://transweb.sjsu.edu/research/How-Ease-Womens-Fear-Transportation-Environments-Case-Studies-and-Best-Practices

North Coast Rape Crisis Team. (n.d.). Retrieved 25 February 2025, from https://www.humboldt.edu/supporting-survivors/educational-resources/statistics

Park, Y., & Garcia, M. (2020). Pedestrian Safety Perception and Urban Street Settings. *International Journal of Sustainable Transportation,* 14(11), 860–871.

Peachey, A. A., & Baller, S. L. (2015). Perceived Built Environment Characteristics of On-Campus and Off-Campus Neighborhoods Associated with Physical Activity of College Students. *Journal of American College Health,* 63(5), 337–342.

Poobalan, A. S., et al. (2012). Physical Activity Attitudes, Intentions and Behaviour Among 18–25 Year Olds: A Mixed Method Study. *BMC Public Health*, 12(1), 640.

Rierson SL. (1994). Race and Gender Discrimination: A Historical Case for Equal Treatment Under the Fourteenth Amendment. *Duke Journal of Gender Law and Policy*, 1, 89–117.

Rustin, S. (2014). If Women Built Cities, What Would Our Urban Landscape Look Like? *The Guardian*. Retrieved 20 December 2023, from https://www.theguardia.com/cities/2014/dec/05/if-women-built-cities-what-would-our-urban-landscapes-look-like

Solsten, E., McClave, D. (1994). *Austria: a country study.* Library of Congress Cataloging-in-Publication Data, ISSN 1057–5294.

Stodolska, M., et al. (2013). "I Was Born in the Hood": Fear of Crime, Outdoor Recreation And Physical Activity Among Mexican-American Urban Adolescents. *Leisure Sciences*, 35(1), 1–15.

Tauberer, J. (2020). What Three Weeks in Japan Taught Me About Public Transit Signage. *Medium*. Retrieved 20 December 2023, from https://joshuatauberer.medium.com/what-three-weeks-in-japan-taught-me-about-public-transit-signage-77bfc2d6e824

Telama, R., et al. (2005). Physical Activity From Childhood to Adulthood: A 21-Year Tracking Study. *American Journal of Preventive Medicine,* 28(3), 267–273.

Chapter 7

Addressing Health, Abilities, Sexuality, Gender, Age, Race, Language, Income, and Culture – Issues, Solutions, Materials, and Design Overview

Images 7.1–7.9 **Realizing how public and private design is not inclusive – as well as intentionally using design for equity to address health, abilities, sexuality, gender, age, race, language, income, and culture – is necessary for the greater cultural growth.**

DOI: 10.4324/9781003475293-8

7.1 Introduction

There are numerous design approaches to address equity and inclusion that have been discussed and illustrated throughout this book. It is important to understand the approaches and green designs presented in earlier chapters, as well as understanding that there are more approaches and understandings. This chapter has lists of ways to understand design issues for health, abilities, sexuality, gender, age, race, language, income, and culture. The next list then includes design and green solutions for these.

The second part of the chapter has more information about green materials and other design terminology that have been discussed in the book. Understanding how green materials contribute to healthy green design toward the greater goal of equity and inclusion is an important goal toward achieving this.

When in spaces where there are subtle and loud ways in which equity and inclusion are not being practiced, take note: report the design issues to the administrating bodies and to the public, as well as proposed changes. Seeing the design issues is a fantastic first step, leading to an even better second step of design solutions that address the problems.

When seeing problems leads toward solutions, people are able to then focus on the solutions to achieve change.

7.2 Issues in Green Design Related to Health, Abilities, Sexuality, Gender, Age, Race, Language, Income, and Cultures, and Creating Design and Green Solutions

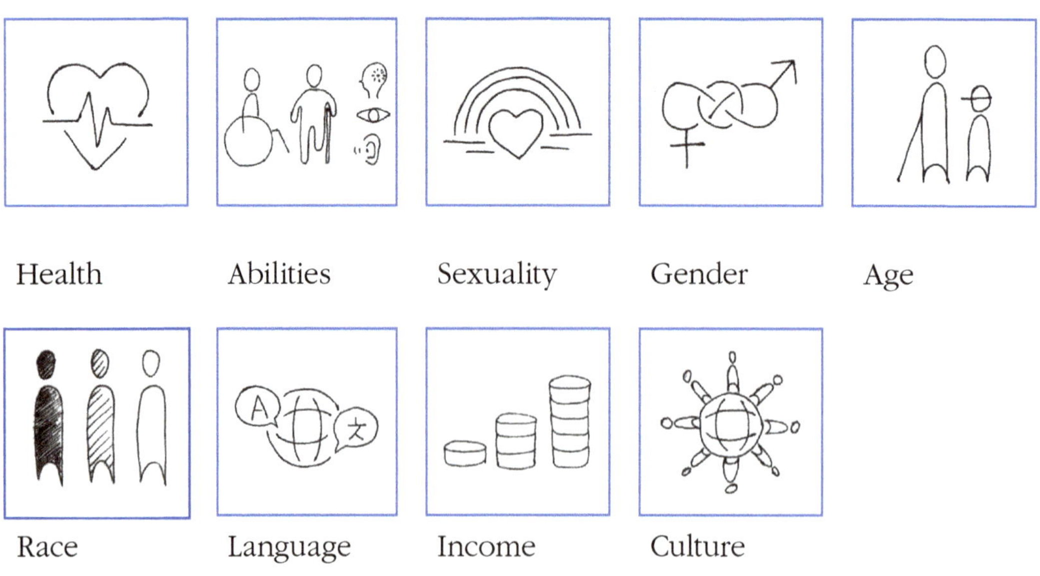

There are a range of issues and solutions addressing design for health, ability, sexuality, gender, age, race, language, income, and culture included in this section. Some of the same items listed on these next pages are applied in the designs of the previous chapters.

The issues in this section for each of the headers are labeled with numbers. *The numbering does not address priorities on the lists.* Rather, the numbering is for ease of reading, as well as to relate to the order of the design solutions in the next section. As with the issues lists, the solutions in this section for each of the headers are numbered. *Again, the numbering does not address priorities on the lists.* Rather, the numbering is for ease of reading.

Existing solutions – as well as unique solutions for urban, suburban, and rural sites – are listed in this section. The next time you are out walking or traveling around sites, consider how the following solutions can apply – or develop more solutions!

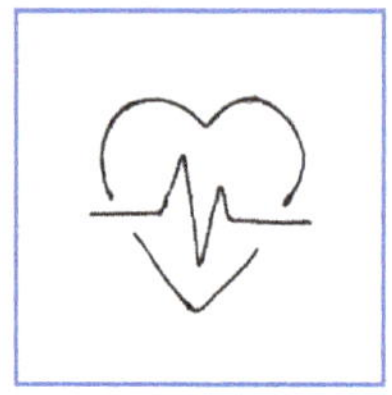

Health design issues (physical and emotional) include the following:

1. Lack of access to local green spaces and green designs for healthy air.
2. Lack of access to clean drinking water.
3. Lack of access to local public green spaces for healthy areas to play and move.
4. Lack of access to local public green spaces for healthy food, including community gardens.
5. Lack of access to local public green spaces for community and social gathering.
6. Lack of access to local public green spaces for reducing stress and improving emotional health.
7. Public spaces and private spaces use materials that off-gas, producing contaminants in the air.
8. Lack of access and views from private and public buildings to green spaces for reducing recovery time from illness.
9. Hospitals and healthcare facilities have poor interior lighting, which is not effective for recovery.
10. Hospitals and healthcare facilities have sounds and smells that are not effective for recovery.
11. Color theory and psychology are not effectively used in selecting materials for emotional healing.

Image 7.10 **Area with flooding affecting the health of the residents.**

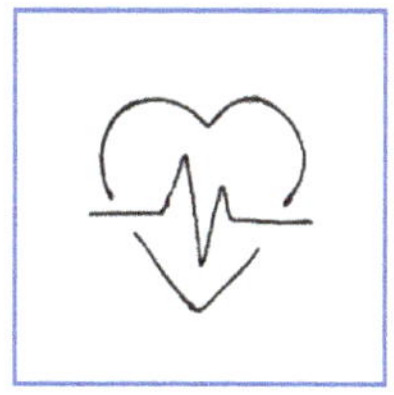

Health design and green solutions (physical and emotional, which are further illustrated and explored in Chapters 2–6) include the following:

1. Create local green spaces for healthy air.
2. Create green designs to improve water quality.
3. Create local public green spaces for playing and moving.
4. Create local public green spaces for healthy food, including community gardens.
5. Create local public green spaces for community and social gatherings.
6. Use materials low in off-gassing to reduce air contaminants.
7. Add comfortable seating in spaces for people with different healing needs, including varying heights and mobility for people with different health needs, including children, mature adults, people with varying weights and mobility, and people in wheelchairs.
8. Create access to local public green spaces for reducing stress and improving emotional health.
9. Create access to and views of local green spaces from private and public windows to reduce recovery time from illness.
10. Create a living design with healing scents.
11. Create a living design with plant lights and relaxing lights.
12. Use lighting in creating playful and illustrative shadows with the design at night.
13. Choose greenery that changes and grows over time for people to experience changes such as flowering or new leaves, which improves healing by seeing plants grow and change in magical ways in a short time.
14. Use outdoor green designs to attract animals for a healthy site and the enjoyable sights and sounds of animals.
15. Use plants that are lighter and more airy, such as those that move in the wind, for visual relaxation and the relaxing sounds of wind through moving plants.

Image 7.11 **Using green design to reduce the effects of flooding, improving the health of the residents.**

16. Create a green design in a context of a bigger view that is enjoyable, which may include creating a mural or other background – something bigger and inspiring.
17. Use colors for joy and healing.
18. Create spaces that feel semi-private from some views and quiet for people to have a sense of semi-privacy and quietness with their thoughts while healing.
19. Add seating for sitting alone, and some for sitting with others for company.
20. Place seating to have inspiring green views.
21. Have spaces with seating in a hospital or health facility that have healing scents, including scented oils, for healthy sensory experiences.
22. Have spaces with views to green design, which may be a living mural, living mobile sculpture, or outdoor spaces.
23. Create mobile green pieces that can be moved around in healthcare facilities for different patients and workers to enjoy living designs.
24. Create walking routes in public spaces, healthcare facilities that loop and have different directions (without dead ends) to take for people to walk and experience enjoyable views.
25. Create spaces in a healthcare facility where people can put on headsets when seated to enjoy unexpected enjoyable sounds, including soothing ocean waves or other healing sounds.

Image 7.12 **Using green design to reduce the effects of flooding, improving the health of the residents.**

26. Use vertical spaces in public sites and healthcare facilities for green walls, living art, or other art that changes and can be enjoyed for an extended time and seen differently at different times of day or times of year.
27. Design spaces that can be interactive for improving emotional healing.
28. Design for a fuller sensory experience of sight, sound, touch, taste, smell, and proprioception (for people to keep track of where their body is in space).
29. Design spaces with multiple entry points and exit points for walking different distances and having different experiences on the walk.
30. Create spaces and green designs for people with different mobility, height, and abilities.

Ability design issues (including people with different physical and learning abilities) include the following:

1. Scenic views are designed to be experienced by people with six-foot standing height, not people who are in wheelchairs or shorter.
2. In the United States, the Americans with Disabilities Act was put into law in 1990. The application of the law to existing sites takes time and is applied as an afterthought. The spirit of the law is for equity; however, the lack of equitable applications of ADA design solutions exists.

Images 7.13–7.14 **QR codes have gained use for people with different visual abilities in options for reading signs through their phone in multiple languages and with updated information.**

3. A lack of applying the spirit of the ADA to create equitable spaces and experiences for inclusion of people with different abilities.
4. In university public design courses, ADA applications are rarely taught.
5. When ADA is taught in design courses, it is typically brief and described as an afterthought or nuisance to the design process, rather than being celebrated for inclusion in the full design process.
6. At design offices, ADA is practiced, though not typically celebrated for a great scope of inclusivity in the design process.
7. Multiple modes of moving through a space are an asset but are not treated that way by numerous designers and policymakers.
8. Lacking inclusivity in design spaces continues the cycle of lacking inclusivity going forward.
9. The creators of the design spaces are not typically people who have different abilities.
10. Further, people with different abilities are typically not included in the design process of spaces.

Ability design and green solutions (including different physical and learning abilities or design, further illustrated and explored in Chapters 2–6) include the following:

1. Create scenic views designed to be experienced by people who are in wheelchairs or are below average height.
2. In the United States, the ADA was put into law in 1990. The application of the law to existing sites takes time, and is applied as an afterthought. The spirit of the law is for equity; however, the lack of equitable applications of ADA design solutions is notable.
3. Applying the spirit of the ADA to create equitable spaces and experiences for inclusion of people with different abilities throughout private and public spaces.
4. Include required courses at universities for ADA design through public design curriculum and beyond.
5. Teach and celebrate ADA in multiple design courses as an inclusive asset in the design process.
6. Celebrate the inclusivity of ADA design at design offices in the design process.
7. Create multiple modes of moving through spaces as an asset.

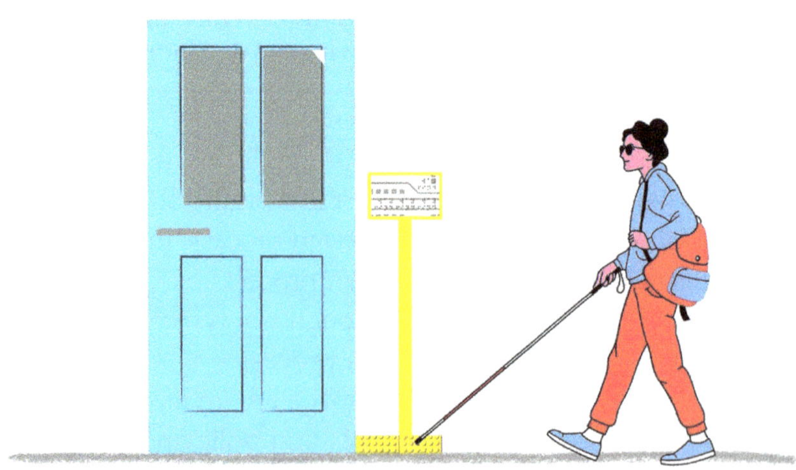

Image 7.15 **Wayfinding through a space with high-contrast sign indicators in a walkway, along with braille and a QR sign, is an effective design approach for people with different visual abilities.**

8. Highlight and publicize effective ADA design in spaces, creating awareness and encouraging future designs.
9. Include people with different abilities in the design process of spaces.
10. Design signage to display symbols, images, braille, and more to be inclusive of people with different visual, hearing, reading, and language abilities and skills.
11. Create universal symbols, icons, and graphics for international use for people to understand and navigate through spaces clearly.
12. Create digital signs that change languages and colors for readability for people with different visual abilities.
13. Create signs with QR codes for people to use to scan and hear the words in a language they understand.
14. Create designs and spaces that acknowledge and celebrate people with different abilities.
15. Create memorials and tributes for people with different abilities in public design.
16. Name and rename public spaces, buildings, structures, and streets for people with different abilities.
17. Create public art for people with different abilities.
18. Create platforms to address the needs of people with different abilities.
19. Design public warm lighting and walkway lighting for addressing concerns of safety related to how people with different abilities feel in spaces.
20. Design wide sidewalks for people using walking support.
21. Repair sidewalks and walkways that have cracks for safety for people using walking support.

Image 7.16 **High-contrast textured surfaces used in design are an effective wayfinding approach for people with different visual abilities.**

22. Design walkways that are easily walkable with surface materials that are safe for people using walking support.
23. Use textured surfaces with distinctive sounds as people walk using a cane for people with different visual abilities to hear as they walk into different spaces.
24. Use surfaces with distinctive textures for people with different visual abilities to feel as they walk into different spaces.
25. Design furniture for a diverse group of people for the inclusion of those with different abilities and needs to make them all comfortable.
26. Create designs and images of the space that fully and sincerely represent inclusive use of the space for womxn, people of global cultures, people who are Black, Brown, and underserved the LGBTQIA+ community, and people of different abilities.
27. Create planted areas to reduce background sounds for people with different hearing to more clearly hear the space they are in.
28. Create crossings with universally understood sound indicators at safe crossing areas.
29. Create crossings with enough time for people with different walking abilities to cross safely.
30. Create safe gaps in walkways for people with walkers, people using tapping canes, and people in wheelchairs.
31. Add an edge between the sidewalk and green spaces of crushed stones or a texture to indicate for people who are visually impaired that the soft surface of grass is near and to help people with a walker or wheelchair to have a brief transitional edge to not get their canes, walkers, or chair wheels stuck in the grass.
32. Have dedicated space off the walkway for trash, seating, signs, bikes, fire hydrants, e-bikes, scooters, and other items for the walkway to be obstacle-free for pedestrians.
33. Design tactile textured strips on the floor below a braille sign for people using a cane to know there is a sign with braille or a QR code that can be scanned with readable options.
34. Create tactile indicators on ramps, steps, and walkways for people who are visually impaired to understand the spaces they are in.
35. Design safety texture strips on the edge of the steps for people who have different visual abilities or are using a cane so that they know the location of the edge of the steps.
36. Create handrails on ramps, steps, and walkways for safer navigation, including those with braille, text, or other ways to improve safe wayfinding.
37. Create handrails for people of a variety of heights.
38. Design handrails and posts that do not cause walking canes to get caught in them.

Image 7.17 **Redesigning staircases with ramps integrated into the design for pushing strollers, or for use by people who use wheelchairs or walkers.**

39. Create ramps and lifts in public spaces for people with different walking abilities.
40. Create clear signage for understanding navigation with clear and distinct colors, symbols, and icons, as well as words.
41. Create universal symbols and colors for people with different visual abilities to have a clear understanding.
42. Use consistent heights, colors, textures, symbols, surfaces, signs, and sounds for clear navigation through spaces.
43. Use clear signage for the routes for people who cannot use stairs due to different walking abilities.
44. Create digital signs for emergency events.
45. Put signs at heights for a variety of people, including people who are shorter than average height or who use wheelchairs.
46. Design signs with large fonts and contrasting colors for clearer readability.
47. Use colors, textures, and symbols on signs for people with autism.
48. Create spaces that reduce background sounds and have textures, and scents for people with autism.
49. Put fire alarms in a standard and safe location away from the path of the push button for opening doors.
50. Design swings and other playground features for use by people in wheelchairs or other different abilities that are also for users who are able-bodied so that children can play together.

Sexuality design issues (LGBTQIA+) include the following:

1. A disproportionately large percentage of designers and policy makers of spaces are heterosexual.
2. A disproportionately low percentage or lack of people who are LGBTQIA+ are involved in the design process.
3. A disproportionately low percentage or lack of acknowledging and celebrating people who are LGBTQIA+ in designed spaces.
4. A disproportionately low percentage or lack of memorials and tributes for people who are LGBTQIA+ in public design.
5. A disproportionately low percentage or lack of naming of public spaces, buildings, structures, or streets for people who are LGBTQIA+.
6. A disproportionately low percentage or lack of public art for people who are LGBTQIA+ successes in their fields.
7. A disproportionately low percentage or lack of platforms to address people who are LGBTQIA+.
8. A disproportionately low percentage or lack of public warm lighting and walkway lighting for addressing concerns of safety people who are LGBTQIA+ feel in spaces.
9. The types of public spaces being created are disproportionately low or lack the desired safety, welcoming design, and activities of people in the LGBTQIA+ communities and their allies.
10. Lacking inclusivity in design spaces continues the cycle of lacking inclusivity going forward.

Image 7.18 **All-All gender restrooms allow people to identify however they choose when using the public restroom.**

Sexuality design and green solutions for LGBTQIA+ (these are further illustrated and explored in Chapters 2–6) include the following:

1. Include people who are LGBTQIA+ as designers of and policy makers for spaces.
2. Include people who are LGBTQIA+ in the design process.
3. Include people who are LGBTQIA+ for acknowledging and celebrating people in designed spaces.
4. Include people who are LGBTQIA+ in memorials and tributes in public design.
5. Include people who are LGBTQIA+ in naming public spaces, buildings, structures, and streets.
6. Celebrate people who are LGBTQIA+ in public art for people who are successes in their fields.
7. Include social platforms for people who are LGBTQIA+ to address their needs.
8. Create public warm lighting and walkway lighting to address concerns about safety in spaces of people who are LGBTQIA+.
9. Create public spaces that are intentionally acknowledging and welcoming for people who are LGBTQIA+.
10. Lacking inclusivity in design spaces continues the cycle of lacking inclusivity going forward.

Image 7.19 **The Stonewall National Monument in New York City is located within a park and is the first U.S. national memorial dedicated to LGBTQIA+ rights events of the past.**

11. Create community design for people who are single and want a community of support.
12. Design spaces that are semi-private for a sense of safety for the LGBTQIA+ community.
13. Design with queer plant communities.
14. Create signs with gender-neutral language.
15. Create greenbelts (areas of green designs in a connected design) to connect a variety of communities of different cultures, races, and incomes.
16. Create designs and images of the space that fully and sincerely represent inclusive use of the space for womxn; people of global cultures; people who are Black, Brown, and underserved communities; the LGBTQIA+ community; and people of different abilities.
17. Remove public art that is erroneous or demeaning, or promotes negative or inappropriate stereotypes of womxn, LGBTQIA+, Black, Brown, and underserved communities.
18. Create inclusion in design spaces to drive inclusivity going forward.

Image 7.20 Inclusive public art that acknowledges and celebrates the LGBTQIA+ community, a marginalized community.

Gender design issues of womxn in designed spaces include the following:

1. A disproportionately large percentage of designers of and policy makers for spaces are men (Zippia, n.d.).
2. A disproportionately low percentage or lack of womxn were allowed to purchase properties due to discrimination for an extended time.
3. A disproportionately low percentage or lack of womxn in the design process.
4. A disproportionately low percentage or lack of acknowledging and celebrating womxn in designed spaces.
5. A disproportionately low percentage or lack of memorials and tributes for womxn in public design.
6. A disproportionately low percentage or lack of naming of public spaces, buildings, structures, and streets for womxn.
7. A disproportionately low percentage or lack of public art for womxn's successes.
8. A disproportionately low percentage or lack of platforms to address womxn's needs.

Image 7.21 **Walkway materials are uncomfortable, unusable, or unsafe for walking in heels or boots, pushing strollers, and walking with a child or multiple children.**

9. A disproportionately low percentage or lack of public warm lighting and walkway lighting for addressing concerns on how womxn feel in spaces.
10. A disproportionately low percentage or lack of public design for womxn who are pregnant.
11. A disproportionately low percentage or lack of semi-private and public spaces for womxn who are nursing.
12. A disproportionately low percentage or lack of safety in public spaces with warm and consistent lighting that addresses the feel of safety for womxn (Boomsma and Steg, 2014; Chapman, n.d.; Cohen et al., 2021; Etopio et al., 2019; Foster et al., 2014; Loewen et al., 1993).
13. A disproportionately low percentage or lack of safety on walkways due to narrow widths of sidewalks. Wide sidewalks are needed for people pushing strollers and walking with others for safety. It is reported that 81% of caregiver are womxn, which includes caregivers of children, spouses, and parents (Sauer, 2022).
14. Sidewalks and walkways that have cracks and are not repaired are unsafe for people wearing heels, pushing strollers, or rolling utility carts during errands.
15. Walkways that are not easily walkable due to surface materials which are unsafe for people wearing heels, pushing strollers, or rolling utility carts during errands.
16. The types of public spaces being created are disproportionately low or lacking in representing the desired activities of womxn. The current types of public spaces created are for sports activities disproportionately

Image 7.22 **Adding handrails to walkways can help womxn who are pregnant and people with different abilities be more able to use the space.**

favorited by men who enjoy cardio activities with groups (e.g., basketball, baseball, football). Activities favorited by womxn include tennis, yoga, meditation, walking, running, and playgrounds (Barnes, 2021; Cohen et al., 2021; Hunt, 2019; Nagendra and Sen, 2022; Rustin, 2014).

17. Womxn make up a larger percentage of users of public transportation in the United States (Goodyear, 2015). There is a disproportionately low percentage or lack of stopping points on routes of public transportation for womxn who are running multiple errands on public transportation.
18. A disproportionate number of public restrooms for womxn, based on the time they spend in the restroom. Men spend an average of 35 seconds at a urinal and 60 seconds overall in the restroom. Womxn who do not have children with them spend an average of 90 seconds in the restroom, as they need to open the stall door, latch the door, attach a bag to a hook, remove layers of clothing, use the restroom, put their clothing back on, unlatch the stall door, open the door, wash their hands, and leave. If they are changing a sanitary napkin, breastfeeding, or changing a diaper, womxn use the restroom more often and require a garbage can. If a child is with a womxn, the average time in the restroom is five to ten minutes. Consequently, womxn require more restrooms in public spaces for their use, as well as more accessible restrooms for womxn with children or those who are caregivers (Woods, 2016). The same number of stalls for womxn and men is not equitable to address the current design practice that favors men.
19. Lacking inclusivity in design spaces continues the cycle of lacking inclusivity going forward.

Gender design and green solutions of womxn in designed spaces (these are further illustrated and explored in Chapters 2–6) include the following:

1. Include a proportional percentage of womxn as designers of and policy makers for spaces.
2. Reparations addressing the short-term and long-term effects of womxn not being allowed to own land and other events of the recent and long past.
3. Include a proportional percentage of womxn in the design process.
4. Create designs and spaces that acknowledge and celebrate womxn.
5. Design memorials, tributes, sculptures, and public art for womxn; those of different abilities; people of global cultures; people who are Black, Brown, and from underserved communities; and the LGBTQIA+ communities.
6. Name and rename public spaces, buildings, structures, and street names for womxn.
7. Create public art that celebrates womxn's successes.
8. Create social platforms to address womxn's needs and listen to them during design.
9. Design public warm lighting and walkway lighting for addressing concerns about safety related to how womxn feel in spaces.
10. Create safe and comfortable public design for womxn who are pregnant and breastfeeding.
11. Create public furniture that is comfortable for people who are pregnant or with children to use in shared spaces.
12. Create semi-private and public spaces for womxn who are nursing.
13. Design wide sidewalks for people pushing strollers, caregivers, and those who are walking with others. As womxn are the majority in this category, creating wider sidewalks improves the quality of life and safety for womxn.
14. Create semi-private spaces with warm lighting that address the sense of safety which womxn want for both daytime and night-time use.
15. Repair sidewalks and walkways that have cracks for safety for people wearing heels, pushing strollers, or rolling utility carts during errands.

16. Design walkways that are easily walkable with surface materials that are safe for people wearing heels, pushing strollers, or rolling utility carts during errands.
17. Create public spaces for the desired activities favored by womxn, including asking womxn what they want, listening to their needs, and following through.
18. Create unisex toilets, which means everybody waits the same amount of time for use.
19. Create a 3:2 ratio of womxn's to men's restrooms, which accounts for the needs of womxn.
20. Womxn need public toilets to be safe. This can be achieved by placing entrances and exits to toilet facilities that allow womxn direct access from public areas, with mirrors strategically located to assist the line of sight and emergency call buttons.
21. Create public spaces that are intentionally acknowledging and welcoming for womxn.
22. Design public transportation that has stopping and resting points along routes for womxn who are running multiple errands, caring for children, and changing routes.
23. Create signage, language, designs, and images that fully and sincerely represent inclusivity for womxn; people of global cultures; and people who are Black, Brown, and from underserved communities, the LGBTQIA+ community, and people of different abilities.
24. Design quiet, well-lit, semi-private spaces in public spaces as resting points for womxn and people who need to use them.

Image 7.23 **Seating is not designed for womxn, including comfort for womxn who are pregnant or caring for one or more children.**

25. Acknowledge and celebrate people who are womxn people of global cultures; and people who are LGBTQIA+, and Black, Brown, and from underserved communities.
26. Remove public art that is erroneous or demeaning, or promotes negative or inappropriate stereotypes of womxn and those who are LGBTQIA+ or from Black, Brown, and underserved communities.
27. Create inclusiveness in designed spaces to drive inclusivity going forward.

Image 7.24 **Comfortable transit seating for womxn who are nursing.**

Age issues in design, including those related to people who are mature, children, and teenagers, include the following:

1. Seats and signage are designed for people of five-foot-eight to six-foot-one height to sit comfortably and read signs at eye level. This is not the average height of womxn, children, teenagers, and mature people.
2. Visibility to see children in public spaces is not a typical focus of public designs.
3. Sites are not typically designed for views and tactile experience for the height of children.
4. Text size on signage is too small for readability for people with maturing or aging eyesight.
5. Font colors on signs are not high contrast enough to be clear and visible for people with maturing or aging eyesight.
6. Crosswalks are not typically timed long enough for people with walkers or children who are being led by their parents to cross the street.
7. Sidewalks and walkways that have cracks and are not repaired are unsafe for people who have young walking needs or mature walking needs.
8. Lack of spaces designed specifically for children, teenagers, and people who are elderly.
9. Handrails on sloped walkways are not consistent for safety and ease of navigating stairs and walkways for people with young or mature walking needs.
10. Use of textured strips and bright colors to aid with safe walking are not consistent on public steps.

Age design and green solutions, including people who are mature, children, and teenagers (these are further illustrated and explored in Chapters 2–6) include the following:

1. Design seats and signage for people who are shorter than six feet tall.
2. Create public furniture that is comfortable for people of mixed ages to use as shared spaces.
3. Design for visibility to see children in public spaces.
4. Design sites for views and tactile experience for the height of children.
5. Use text size on signage that is readable for people with aging eyes.
6. Use font colors on signs that are high contrast enough to be clear and visible.
7. Design timers on crosswalks with a comfortable amount of time for people with walkers or children who are being led by an adult to cross the street.
8. Repair sidewalks and walkways that have cracks to make them safe.
9. Create spaces designed specifically for children, teenagers, and people who are mature.

Image 7.25 **Create comfortable seating for mature adults to enjoy public spaces.**

10. Create designs that more fully and sincerely represent inclusive use of the space for a range of ages.
11. Design handrails on sloped walkways that are consistent for safety and ease of navigating stairs and walkways.
12. Use textured strips and bright colors on steps to aid with safe walking on public steps.

Image 7.26 **Create shared spaces for mature adults, teenagers, and young children to be active together.**

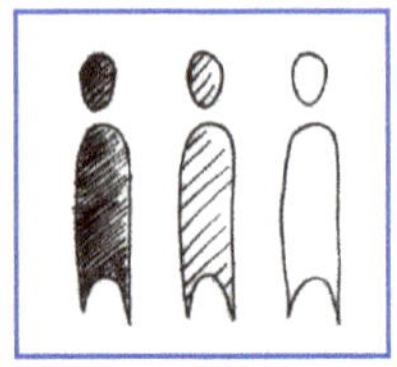

Race design issues affecting people who are Black, Brown, Indigenous, mixed race, and from underserved communities include the following:

1. A disproportionately low percentage or lack of people who are Black, Brown, and from underserved communities were allowed to purchase properties due to discrimination for an extended time.
2. The short-term and long-term effects of people who are Black, Brown, and from underserved communities not owning property create a lack of ability to build wealth and create generational wealth.
3. A disproportionately low percentage or lack of designers who are Black, Brown, and from underserved communities.
4. A disproportionately low percentage or lack of people who are Black, Brown, and from underserved communities in the design process.
5. A disproportionately low percentage or lack of acknowledging and celebrating people who are Black, Brown, and from underserved communities in designed spaces.
6. A disproportionately low percentage or lack of memorials and tributes for people who are Black, Brown, and from underserved communities in public design.
7. A disproportionately low percentage or lack of naming of public spaces, sites, buildings, structures, streets and more for people who are Black, Brown, and from underserved communities.
8. A disproportionately low percentage or lack of acceptance of signs, flags, symbols, and designs for people who are Black, Brown, and from underserved communities.
9. A disproportionately low percentage or lack of public art celebrating the successes of people who are Black, Brown, and from underserved communities.
10. A disproportionately low percentage or lack of acceptance of ways in which people who are Black, Brown, and from underserved communities are allowed to celebrate their race (e.g., graffiti as art form); it is more regulated for people who are Black, Brown, and from underserved communities.
11. A disproportionately low percentage or lack of platforms to address the needs of people who are Black, Brown, and from underserved communities.

12. A disproportionately low percentage or lack of acceptance of the events of the recent and long past of people who are Black, Brown, and from underserved communities.
13. Underserved communities have limited spaces where people can live, and often these are in areas with poor air quality and healthy site issues.
14. Underserved communities have limited spaces where people can play and gather.

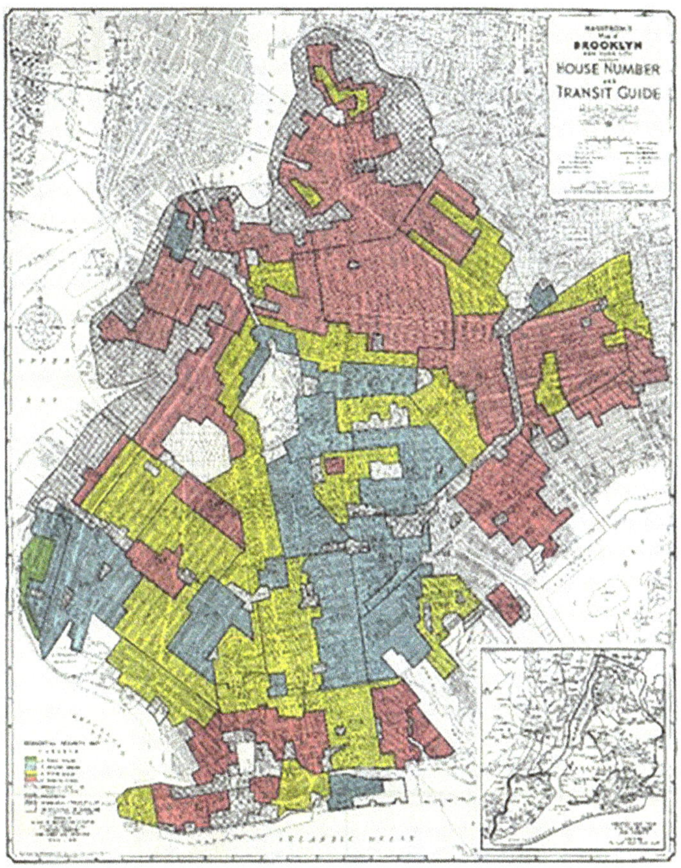

Image 7.27 **Redlined maps in Brooklyn, New York, and throughout the United States illustrate areas where people who are Black were unable to live or purchase property.**

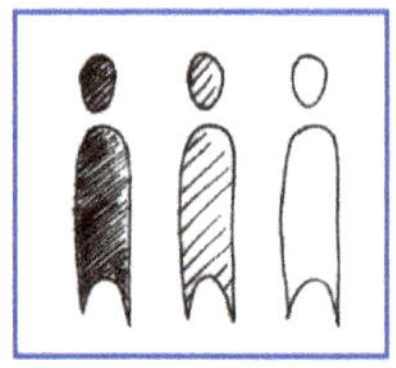

Race design and green solutions related to people who are Black, Brown, Indigenous, mixed race, and from underserved communities (these are further illustrated and explored in Chapters 2–6) include the following:

1. Reparations addressing the short-term and long-term effects of people who are Black, Brown, and from underserved communities not being allowed to own land and other events of the recent and long pasts.
2. Intentional inclusion to celebrate the events of people who are Black, Brown, and from underserved communities.
3. Intentional inclusion of people who are Black, Brown, and from underserved communities as designers of and policy makers for public spaces.

Image 7.28 **Public memorials, including this one for George Floyd, acknowledge issues of racial inequity. George Floyd was murdered in Minneapolis, Minnesota, on 25 May 2020 by a police officer kneeling on Floyd's neck for eight minutes and 46 seconds. That police officer was sentenced to prison after being found guilty of murder as well as violating Floyd's civil rights.**

4. Intentional inclusion of people who are Black, Brown, and from underserved communities in the design process.
5. Intentional inclusion of people who are Black, Brown, and from underserved communities in memorials and tributes in public design.
6. Intentional inclusion of people who are Black, Brown, and from underserved communities of naming of public spaces, sites, buildings, structures, streets, and more.
7. Intentional inclusion of people who are Black, Brown, and from underserved communities in acceptance of signs, flags, symbols, and designs.
8. Intentional inclusion of public art in diverse forms inclusive of people who are Black, Brown, and from underserved communities.
9. Remove strict regulations on public art such as graffiti, which celebrates people who are Black, Brown, and from underserved communities.
10. Design memorials, sculptures, and public art for womxn; people of global cultures; people who are Black, Brown, and from underserved communities; and LGBTQIA+ communities.
11. Create designs and images of the space that fully and sincerely represent inclusive use of the space for womxn; people of global cultures;

Image 7.29 **The African Memorial in New York City is designed to acknowledge a site that was a cemetery for people who were enslaved.**

people who are Black, Brown, and from underserved communities; the LGBTQIA+ community; and people of different abilities.

12. Create and listen to social platforms designed to address people who are Black, Brown, and from underserved communities.
13. Name and rename sites and events for people who are Black, Brown, and from underserved communities.
14. Create signs, flags, symbols, and designs for people who are Black, Brown, and from underserved communities.
15. Create public art about and by people who are Black, Brown, and from underserved communities.
16. Remove public art that is erroneous or demeaning, or promotes negative or inappropriate stereotypes toward womxn; LGBTQIA+; and Black, Brown, and from underserved communities.
17. Create unique living green retrofitted community spaces in underserved communities to improve air quality and health.
18. Create greenbelts (areas of green designs in a connected design) to connect a variety of communities of different cultures, races, and incomes.

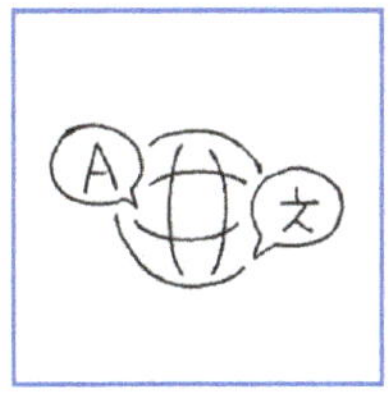

Language design issues, including those affecting people who understand different languages or have different reading abilities, include the following:

1. A disproportionately low number of designers and policy makers are multi-lingual.
2. A disproportionately low percentage of diverse languages are used in signage in public spaces, buildings, structures, street names, and more.
3. In the United States, spaces, buildings, structures, street names, and more are named and re-named in English rather than in their original language. This means that some of the original names in Indigenous or other languages are not being celebrated or are being lost entirely.
4. Areas have people who speak multiple languages, some of which are not used on public and private signs.
5. Nations across the globe have a variety of official languages, as well as changing languages, and some areas without a national language. With changing words and dialogues, generations of people have a different use and understanding of words.

Image 7.30 **The use of symbols for spaces allows people to understand wayfinding, even when words are not in their known language.**

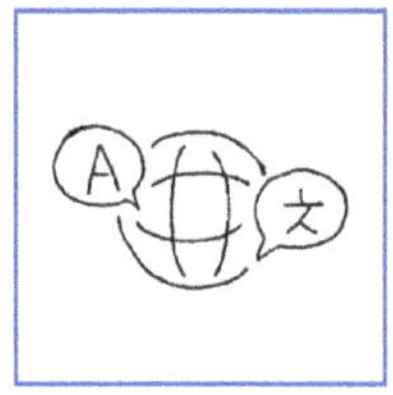

Language design and green solutions, including people who understand different languages or have different reading abilities (these are further illustrated and explored in Chapters 2–6), include the following:

1. Include designers and policy makers who are multi-lingual and understand the differences and nuisances in language communications.
2. Use diverse languages in signage in public spaces, buildings, structures, street names, and more.
3. Name and rename spaces, buildings, structures, streets, and more using Indigenous languages, lost languages, and global languages.
4. Design signage to use symbols, images, braille, and more to be inclusive of people with different reading skills and language skills.
5. Create universal symbols, icons, and graphics for international use for people to understand and navigate through spaces clearly.
6. Use contrasting colors for clarity and readability.
7. Create digital signs that change languages.
8. Create signs with QR codes for people to use to find a language they understand.

Image 7.31 **The use of symbols for spaces allows people to understand wayfinding, even when words are not in their known language.**

Income design issues, for people who have low income, include the following:

1. A disproportionately low percentage or lack of local green spaces for social gathering and playing in low-income neighborhoods and communities (when compared with middle-income and high-income neighborhoods and communities) (Pool et al., 2020; Yang et al., 2017).
2. A lack of public green spaces in low-income neighborhoods negatively affects air quality and emotional health for people in the community.
3. A lack of access to green parks, green public spaces, and green roofs in low-income areas. These spaces improve air quality and emotional health for people in the community.
4. A disproportionately low percentage or lack of public transportation in low-income neighborhoods and communities.
5. People need vitamin D for health and to fight depression. This is produced when people spend time in sunlight. When there is a lack of public green spaces, people do not spend as much time outdoors; therefore, they are not producing vitamin D.
6. A disproportionately low percentage or lack of healthy food, community gardens, and agriculture (called food deserts) in low-income neighborhoods.

Image 7.32 **This low-income area has little green space on the individual properties and in public spaces.**

7. Areas that are formerly redlined sections of cities have a disproportionately low percentage or lack of green spaces for community gathering.
8. Areas that are formerly redlined sections of cities have a disproportionately low percentage or lack of property ownership by families, furthering the problems of a lack of generational wealth still showing today (Gjording, 2022; Nardone et al., 2021; Public Health Columbia University, 2021; Ward, 2021).
9. Industrial companies are typically nearby or within low-income neighborhoods, poorly affecting the air quality, water quality, and soil quality of these communities.
10. The lack of trees and plants in low-income areas allows sounds to bounce and sound louder. Greenery is one way to reduce sounds.
11. Areas that are prone to flooding have been areas where state administration develops for people with low income and underserved communities (Mittan, 2020; Scrader, 2023; Takin et al., 2023).

Income design and green solutions for people who have low income (these are further illustrated and explored in Chapters 2–6) include the following:

1. Design local green spaces for social gathering in low-income neighborhoods and communities.
2. Create new and retrofitted green designs in low-income neighborhoods to improve air quality and emotional health.
3. Create green design public transportation in low-income neighborhoods and communities.
4. Add bike lanes for different modes of transportation.
5. Create green design walk routes for people in low-income communities.
6. Create access to local green spaces, including green street buffers, street trees, expanded street trees, green walls, and unique site-specific greening for healthy air.
7. Re-route trucking through low-income neighborhoods to other areas to reduce noise, air pollution, and increase safety.
8. Design local green public spaces for people to spend time outdoors, contributing to producing vitamin D from sunlight for health.
9. Design and subsidize local public green spaces, both interior and exterior, for growing healthy food, particularly in food deserts.
10. Use street trees, street medians, and other unique sites to plant fruits and vegetables, producing food for people who are low income to have access to unique food sources.
11. Design green areas to reduce the noise pollution produced by nearby industrial companies or other sources.
12. Create green designs to improve air quality, water quality, and soil quality.
13. Use green infrastructure and green design on site, neighborhood, and community scales to reduce the effects of heavy storms and flooding.
14. Create greenbelts (areas of green designs in a connected design) to connect a variety of communities of different cultures, races, and low incomes.
15. Create vertical greening on building sides for unique living murals for people to experience from their windows, improving emotional health with the growing mural, sounds of birds, and improved air quality.
16. Create and enforce policies and procedures that require low-income areas have designated green spaces. Further, require that the spaces stay green and and affordable for people with low income, without gentrification and over-pricing.

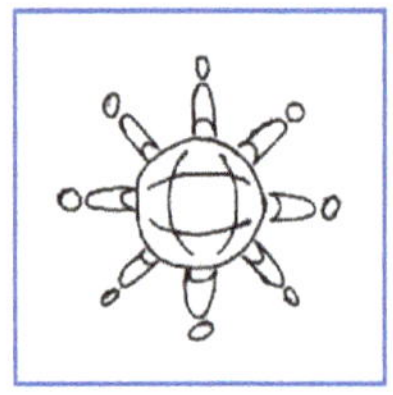

Culture design issues for people of cultures from around the globe and mixed cultures, include the following:

1. A disproportionately low percentage or lack of acceptance of ways in which people of global cultures are allowed to celebrate their culture; it is more regulated for people of global cultures.
2. A disproportionately low percentage or lack of naming of sites and events for inclusion of people of global cultures.
3. A disproportionately low percentage or lack of acceptance of signs, flags, symbols, and designs for people of global cultures.
4. A disproportionately low percentage or lack of public art by and representing people of global cultures.
5. A disproportionately low percentage or lack of representation of people of global cultures in policy makers and designers of public spaces (Zippia, n.d.).
6. A disproportionately low percentage or lack of acceptance of the events of the recent and long pasts of people of global cultures.
7. Allowable use of public and private spaces and types of businesses that are inequitable for people of different cultures.
8. Agriculture that is not inclusive of diverse cultures and immigrants in some areas.
9. A lack of plants that are inclusive of diverse cultures and immigrants.
10. A lack of designs of public spaces that acknowledge and celebrate diverse cultures.

Image 7.33 **Adding green design to low-income housing as a right for people for their emotional and physical health.**

Images 7.34–7.37 **Designs of space from different cultures celebrate people's past, present, and future. Architecture of India, Rome, Great Britan, and China.**

Culture design and green solutions for people of cultures from around the globe and mixed cultures (these are further illustrated and explored in Chapters 2–6) include the following:

1. Create designs that are accepting and celebratory of diverse cultures.
2. Name sites and events for people of global cultures.
3. Design signs, flags, and symbols for people of global cultures.
4. Design public art by and of people of global cultures.
5. The inclusion of people of diverse cultures in policy making and creation of public spaces.
6. Acknowledge and address events of the recent and long past of people of global cultures.
7. Inclusion of public and private spaces and types of businesses for diversity of cultures.
8. Create local and regional agriculture that is inclusive of diverse cultures and immigrants.
9. Design with plants that are inclusive of diverse cultures and immigrants.
10. Create designs that acknowledge and celebrate diverse cultures.
11. Create greenbelts (areas of green designs in a connected design) to connect a variety of communities of different cultures, races, and incomes.

Image 7.38 **Mural of Diverse Faces in Edgewater Neighborhood of Chicago, Illinois, celebrates culture through public art.**

12. Design memorials, sculptures, and public art for womxn; people of global cultures; people who are Black, Brown, and from underserved communities; the LGBTQIA+ community; and people with different abilities.
13. Create designs and images that are fully and sincerely represent inclusive use of the space for womxn; people of global cultures; people who are Black, Brown, and from underserved communities; the LGBTQIA+ community; and people of different abilities.

Image 7.39 **The archway of Chinatown at San Francisco, California, celebrates culture in public design.**

Other design issues that do not fall into the previous categories include the following:

1. Access to local public space is treated as a luxury rather than as a right.
2. Safe and social green spaces are treated as luxuries rather than as rights.
3. The needs and wants for safety in design is different for different people – including womxn, people who are LGBTQIA+, children, people who are mature, and people with different abilities.
4. Physical and psychological safety is not typically designed in public spaces for the needs and wants of womxn, the LGBTQIA+ community, or children, leaving people not wanting to be in spaces that do not feel safe, or not going out to some public spaces.
5. Designing for comfort and ease of walkers is not prioritized over vehicles and public transportation in some areas.
6. Accessibility for pedestrians in design is not prioritized.
7. Funding for projects, including public projects, is given by individuals and companies who prefer to give money to projects that may not include equitable designs.

Image 7.40 **Gramercy Park in New York City is a gated park and green space. This illustrates that green space is considered a luxury rather than a right in public spaces.**

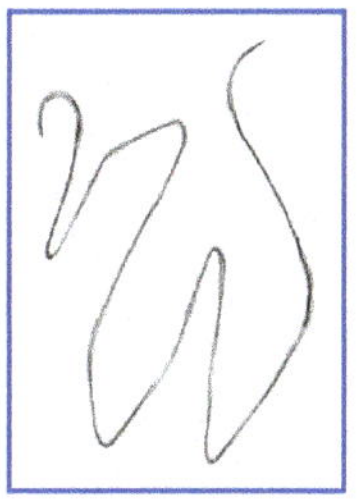

Other design and green solutions that do not fall into the previous categories (these are further illustrated and explored in Chapters 2–6) include the following:

1. Make access to local public space a right rather than a luxury.
2. Make safe and social green public space a right rather than a luxury.
3. Create a design process that listens to and addresses the needs and wants for safety in design, which is different for different people – including womxn, people who are LGBTQIA+, children, people who are mature, and people with different abilities.
4. Design for physical and psychological safety in public spaces for the needs and wants of womxn, the LGBTQIA+ community, children, people who are mature, and people with different abilities so that people will want to be in public spaces and feel safe.
5. Create designs that are for comfort and ease of walkers over vehicles and public transportation.

Image 7.41 **Using living design for health, safety, and social gathering supports green design as a right rather than a luxury.**

6. Design for accessibility for pedestrians in designs.
7. Create funding sources for projects that require equitable designs for diverse people.
8. Add green infrastructure and plant nature-based solutions to improve the safety of areas that are prone to damage from stormwater.
9. Add green infrastructure and plant nature-based solutions to improve the safety of areas that are prone to flooding.
10. Add green infrastructure and plant nature-based solutions to rebuild areas that have flooded.

7.3 Overview of Green Design, Terminology, and Materials

7.3.1 Air-Purifying Paints

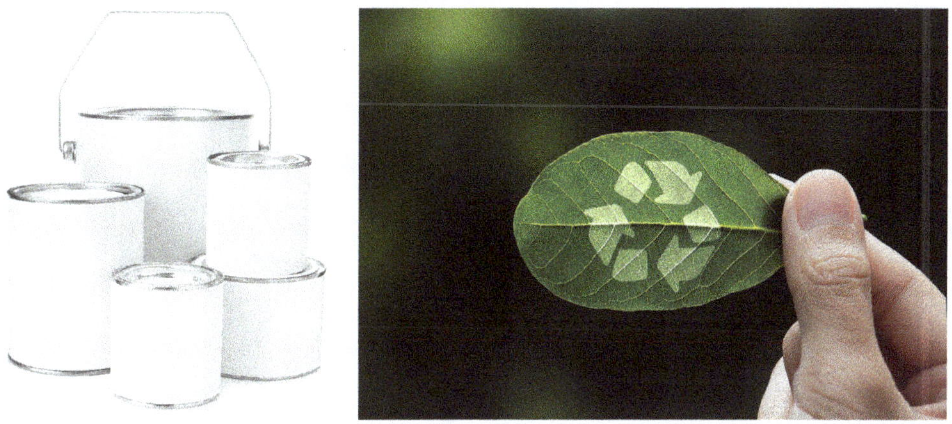

Images 7.42–7.43 **There are a variety of companies that produce natural and air-purifying paints.**

Air-purifying paint is also known as Airlite, Auro, Bona, Ecos paint, Gush, Premir Clean, Super paint, and other companies and brands. A variety of materials are used to reduce air pollutants. An overall understanding is that air-purifying paints improve air quality.

Some of the paints use titanium dioxide (TiO_2, a photocatalytic oxidizer), which is capable of degrading odors, reducing carbon dioxide (CO_2), formaldehyde, and methyl mercaptan, and breaking down VOCs with ultraviolet (UV) light. Some air-purifying paints are natural paints that are made from lime or are predominantly lime-based. When the breathable air-purifying paints are curing, they absorb as much as 4.5kg of CO_2 per 15L of paint due to the natural process of lime curing in the air. Further, air-purifying paints can reduce cigarette smoke, kill bacteria and other micro-organisms, and protect from some molds.

As different companies use their own mixtures and testing, what is important to know is that these paints are out there and improve air quality. While interior paint is the primary selection noted in this section, there is also current research on air-purifying exterior paint. The need for addressing air quality through a myriad of materials and techniques that are best suited for the specificity of the site.

7.3.2 Air Pollutants and Understanding Air Quality of the Site

(Famulari, 2020)

The U.S. Environmental Protection Agency (EPA) has standards for the amount and type of contaminants allowed at acceptable levels. One of the goals for these design solutions is to clean the air to at least the EPA standard, or even cleaner with the use of plant materials.

Excerpts from research by Federico Karagulian et al., 2015, titled "Contributions to Cities' Ambient Particulate Matter (PM): A Systematic Review of Local Source Contributions at Global Level" helps to illuminate air quality in urban areas. The research covers the time period between 1990 and 2014.

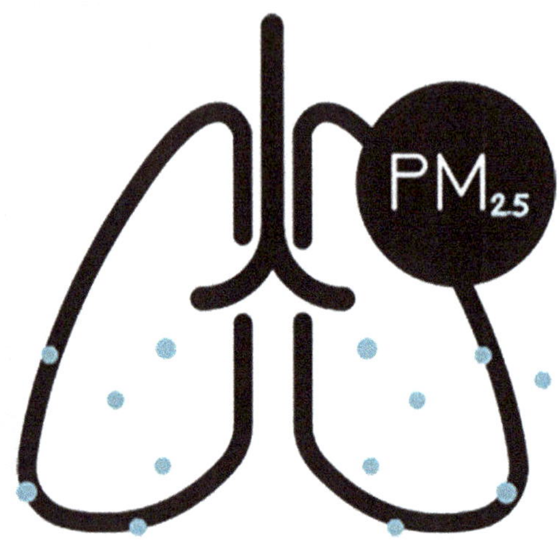

Images 7.44 The particulate matter (PM) in air pollutants is small enough to get into the lungs of anyone who is breathing the polluted air.

There are five categories of main sources of ambient PM commonly found, which have been used for the purpose of this analysis: traffic; industry; domestic fuel burning; natural sources, including soil dust (re-suspended) and sea salt; and unspecified sources of pollution of human origin.

- **Traffic** includes different kinds of emissions from various vehicle types. In addition to primary PM emissions from exhaust, and the emissions of organic and inorganic gaseous PM precursors from the combustion of fuels and lubricants, vehicles emit significant amounts of particles through the wear of brake linings, clutches, and tires (Amato et al., 2009; Belis et al., 2013). These are deposited onto the road and then re-suspended by vehicle traffic, together with crustal/mineral dust particles and road wear material.
- **Industry** is a category including mainly emissions from oil combustion, coal burning in power plants, emissions from different types of industries (petrochemical, metallurgic, ceramic, pharmaceutical, information technology [IT] hardware, etc.) and from harbor-related activities. Industrial sources are sometimes mixed with unidentified combustion sources or traffic (Belis et al., 2013).

- **Domestic fuel burning** includes wood, coal, and gas fuel for cooking or heating, such as in residential complexes.
- **Natural sources** include soil dust and sea salt dust. These components of PM are associated with the re-suspension from fields or bare soils by local winds. When reported separately, road dust was included in the traffic source category in this study (Belis et al., 2013).
- **Unspecified sources of human origin** mainly include secondary particles formed from unspecified pollution sources of human origin. Primary particle emissions include mechanically generated particles and primary carbonaceous particles. Primary particles also include carbonaceous fly-ash particles produced from high-temperature combustion of fossil fuels in coal power plants.

Additional PM is formed from reactions of gaseous pollutants, VOCs, NMVOCs (non-methane volatile organic compounds), and evaporation of solvents such as paints, degreasers, and stain removers.

> Airborne particles like dust, soot and smoke that are less than 2.5 micrometers in diameter are small enough to lodge themselves deep in the lungs. Studies have linked pollution of this sort to respiratory problems, decreased lung function, nonfatal heart attacks and aggravated asthma, according to the United States Environmental Protection Agency.
>
> (Fernandez, 2006)

According to the New York State Department of Health and Department of Environmental Conservation,

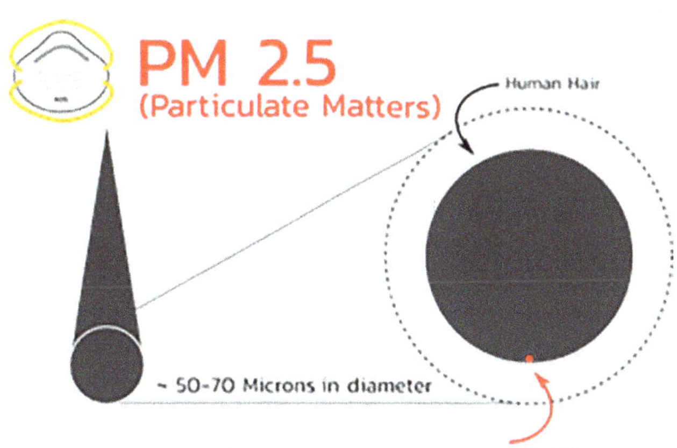

Images 7.45 **Particulate matter 2.5, more commonly known as PM 2.5, is imperceptible and small at approximately 2.5 microns in size. In reference to the diameter of the width of the hair, the small red dot is the size of 2.5 microns.**

Particles in the PM 2.5 size range are able to travel deeply into the respiratory tract, reaching the lungs. Exposure to fine particles can cause short-term health effects such as eye, nose, throat and lung irritation; coughing; sneezing; runny nose; and shortness of breath. Exposure to fine particles can also affect lung function and worsen medical conditions such as asthma and heart disease. Additionally, EPA officials said these fine particles, a significant portion of which are produced by diesel engine emissions, lead to 15,000 premature deaths a year nationwide.

To understand the scale of 2.5 micrometers, there are 25,000 microns in an inch. The width of PM 2.5 is approximately 30 times smaller than that of a human hair. PM 2.5 is quite small, such that several thousand particles of that size would fit in the period at the end of this sentence.

In addition to necessary policy changes addressing gaseous and chemical release into the air, this chapter explores using plants to remediate for PM in air. Some plants are able to take in the contaminants through their roots, stems, or leaves and continue to live, removing the PM from the air so that humans and animals can breathe in cleaner air with less PM.

Image 7.46 **Algae biopanels, started in 2013 at the Bio-Intelligent Quotient (BIQ) House in Hamburg, Germany, uses microalgae to produce electricity, clean oxygen, and a sludgy biomass that can be processed into combustion fuel, fertilizer, and other uses.**

7.4 Algae Biopanels

Algae biopanels are new sustainable structures that reduce CO_2 and create biomass, energy, and oxygen. These panels go by multiple names, including algae bioskin panels and microalgae bioreactor building panels.

Algae biopanels are put on exterior sites and use flat algae tanks with the addition of sunlight to produce numerous benefits. One benefit is that it takes in CO_2 (which people produce when breathing). With sunlight and photosynthesis, the algae also produce biomass, which generates fresh oxygen. The biomass produced from the algae can be used as fuel, fertilizer, and other useful substances. The more sunlight that is available, the faster the algae grows and the more the processes continue.

The panels also produce shade, which may reduce cooling costs in the summer. When the algae biopanels are structured on the outside of the building, they trap heat in the algae-filled water and produce shade, therefore reducing the cooling cost of the building. The heat and the biomass can also be harvested and reprocessed into biofuel. The biofuel can be fed into the burner that runs the building's hot water system.

There are multiple companies that produce algae biopanels. Companies and sites vary on the exact calculations for energy production. A general calculation is two pounds of CO_2 for each pound of algae in the panel.

Images 7.47–7.48 **Algae biopanels use photobioreactors (PBRs) in a flat plate form. Other forms of PBRs include vertical and horizontal tubes as well as bubble columns.**

7.4.1 Greywater

(Famulari, 2024)

The terms “grey water” and “greywater” are used interchangeably. This water can be used and is currently used for irrigation of planted areas. Greywater comes from sources with mild impurities and includes water that has been used in sinks, showers, and washing machines. Mild soaps in water are not harmful to plants in small quantities. Toilets are not included in this category, as their contaminants are described as “blackwater” with more pollutants that damage plants.

Greywater can be stored in units and then filtered through commercial filters as needed for use. Commercial filters can be used in non-winter temperatures to remove larger particles or contaminants before the water is used for irrigation. The filters help the irrigation tubes not be blocked with larger particles that may be in the greywater.

Greywater systems and filters are cleaned out and flushed of water before winter and restarted in the warmer temperatures of spring. During freezing temperatures, outdoor water systems cannot be used so that water does not freeze within the system and damage it.

The re-use of greywater for irrigation, rather than fresh water, is beneficial when considering the larger benefit of sustainable practices.

Image 7.49 **Signage showing the use of reclaimed water for irrigation of a site.**

Images 7.50 **The use of reclaimed water for irrigation conserves fresh water for necessary uses such as drinking. The practice of greywater irrigation is beneficial for sustainable practices.**

7.4.2 Permeable Pavers

(Famulari, 2024)

Permeable pavers have spaces for water to flow through the stones. Water on the pavers goes back into the ground and water table within the site or is used for growing plants between the pavers. In these spaces, low-growing plant ground covers such as grass and moss thrive. There are a variety of styles of these pavers that can be bought commercially such as those in Image 7.51–7.58, or people can create their own unique designs.

Images 7.51–7.58 Examples of permeable pavers – pavers with spaces where plants can grow – in a variety of styles and patterns.

7.4.3 Psychological Effects of Green Design

(Famulari, 2020)

Another item to explore in design is the psychological benefits of views, green designs, and other designs that connect people to green nature. The advantages of green spaces not only provide ecological benefits negating urban heat, offsetting greenhouse gas emission, and offsetting stormwater; there are also health benefits of plant materials. From the University of Washington's *Green Cities: Good Health*, further mental benefits of green design include the following:

- The experience of nature helps to restore the mind from the emotional fatigue of work or studies, contributing to improved work operations and satisfaction.
- Urban nature, when provided as parks and walkways and incorporated into building design, provides calming and inspiring spaces and encourages learning, inquisitiveness, and alertness.
- Green spaces provide necessary places and opportunities for physical activity. Exercise improves cognitive function, learning, and memory.
- Outdoor activities can help alleviate symptoms of Alzheimer's, dementia, stress, and depression, and improve cognitive function in those recently diagnosed with breast cancer.
- Contact with nature helps children to develop cognitive, emotional, and behavioral connections to their nearby social and biophysical environments. Nature experiences are important for encouraging imagination and creativity, cognitive and intellectual development, and social relationships.

Image 7.59 **The view of a Japanese garden from the interior spaces has similar psychological benefits to the being in the exterior garden itself.**

- Symptoms of attention deficit hyperactivity disorder (ADHD) in children can be reduced through activity in green settings, thus "green time" can act as an effective supplement to traditional medicinal and behavioral treatments.

Further information on the studies stated here can be found in the Bibliography. From the World Health Organization's *Urban Green Spaces and Health* – a review of the evidence, their findings summarize that the available evidence of the beneficial effects of urban green spaces, such as improved mental health; reduced cardiovascular morbidity and mortality, obesity and risk of type 2 diabetes; and improved pregnancy outcomes. Mechanisms leading to these health benefits include psychological relaxation and stress alleviation, increased physical activity, and reduced exposure to air pollutants, noise, and excess heat.

7.4.4 Sound Waves and Using Green Design to Reduce Noise Pollution

(Famulari, 2020)

The next item to address is sound. Urban sound is measured in dB(A), which stands for decibels with an A-weighting. This A-weighting includes an adjustment that takes into account varying sensitivity of the human ear to different frequencies of sound. A-weighted measurements consider the loudness, annoyance factor, and stress-inducing capability of noises with low frequencies and moderate or high volumes. This also takes into account the very low and the very high frequencies and volumes of noise.

Why is this worth understanding? Understanding that sounds near airports, expressways, highways, freeways, trains, and subways have a higher dB(A) allows one to consider the next step of how to use design to lower the dB(A) for those who live or work in these areas.

Published research by Maarten Hornikx (2016) illustrates that plants can reduce the dB(A) ranging from 4–7.5 dB(A). Dense foliage absorbs frequencies, while hard surfaces scatter them. Consider a sound frequency hitting a hard wall, or tree trunk – the sound waves scatter. However, with the softer dense surface of foliage, the frequencies are absorbed. The article states that:

> Vegetation in urban areas has a range of ecological advantages. Vegetation acts as noise reducing possibility in inner city environments, in particular building envelope greening measures. Also, there is growing evidence that visibility of vegetation by itself affects noise perception positively. For inner city environments,

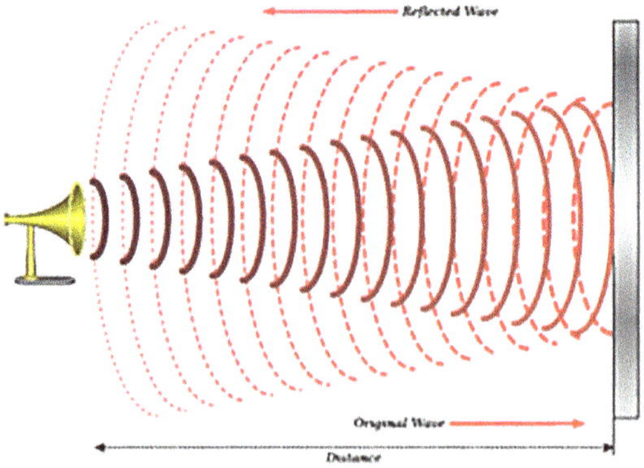

Image 7.60 **Sound coming from a source is directed toward a wall. When the sound waves reach the hard surface of the wall without interference from surfaces in their way, the sound waves are reflected back into the space. The effect can be loud and echoing.**

> vegetation types that can be considered are low-height noise barriers, vegetated façades, vegetated roofs and trees. Low-height vegetated noise barriers were shown to be useful in road traffic noise applications at street level. These devices can be placed close to the driving lanes, thereby yielding road traffic noise reduction of about 5 dB(A).

Some municipalities, such as New York City, require trees as part of their development in order to assist with noise, views, and air quality. Planting trees in New York City sidewalks has been the project of artists such as Alan Sonfist and Joseph Beuys. Even today, walking through New York City, these living artworks can be unrecognized by the community – yet, without these green artworks, the urban area would have deteriorated air quality and lack shade. Other benefits of street trees for a community include experiencing seasonal change, growth, and time changes through plant materials, as well as bringing people together to care for the living plant materials.

In looking at green wall surfaces, it was found that the green wall that does not have plant materials on it only reduces 1 dB(A), while a green wall with plant materials reduces 5 dB(A). It is the soft foliage that absorbs sound.

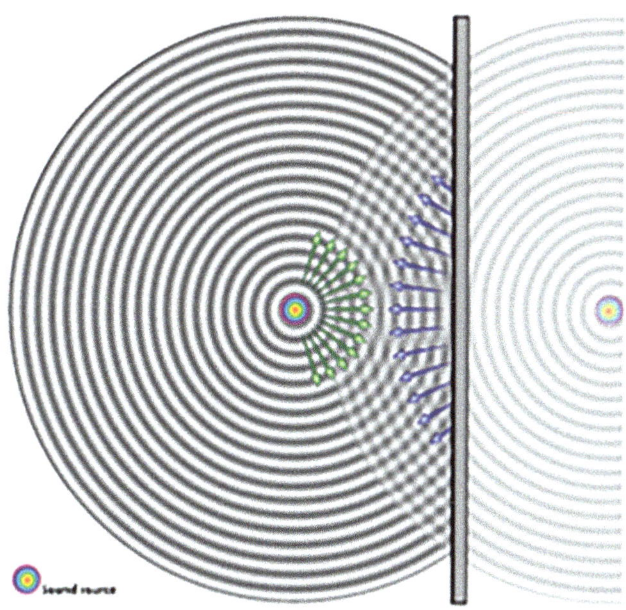

Image 7.61 **Sound waves go in all directions from the source of the sound. When it reaches a hard surface such as a wall, the sound reflects back into the space. Soft surfaces, surfaces with air holes or gaps, and groups of people with space in between absorb parts of the sound wave and reduce it before it gets to the hard wall surface.**

Green roofs can decrease sound waves over buildings. It is the combination of plants and the sublayers of soil and liners, as well as roof shape that produce the benefits of sound reduction. It has been found that buildings with vegetated roofs can have an effect up to 7.5 dB(A) noise reduction.

> The effect of green-wall systems is larger for roadside courtyards than for trafficked streets and may amount up to 4 dB(A). An interesting application of vegetated surfaces are openings to courtyards, greening those surfaces has shown to amount to 4–5 dB(A) reduced sound pressure levels in the courtyard. (p. 103)

What does 5 dB(A) really sound like in volume and effects? Is it really significant?

Yes! An increase of 10 dB(A) feels twice as loud to the human ear. For example, within comparing a dishwasher that runs at 50 dB(A) to a dishwasher that runs at 60 dB(A), the second dishwasher sounds twice as loud at the first: 10 dB(A) feels twice as loud. Therefore, a decrease in dB(A) using plants and green design by 8 dB(A) feels almost half as loud.

The result is the more green, the lower the dB(A), which reduces hearing loss and stress, while creating healthy spaces in an urban area only creates positive outcomes.

7.4.5 Plants Can Absorb Sound Waves

(Famulari, 2020)

Plants can absorb sound through their leaves and foliage. Dense trees are more effective than trees with large spaces without greenery. Barks of trees are hard surfaces that sound waves bounce off and scatter. The green areas of plants are soft and have pockets of air. (Leafier plants are more beneficial for reducing sound waves.) When sound waves are directed to the foliage, they are softened and absorbed into the leaves and air pockets, similar to a soft sponge. This is effective for urban areas where there are a myriad of sounds that move through the air. When dense areas of plant materials are in front of buildings, properties, courtyards, yards, entrances, or windows, the plants act as sponges for sound waves. These plantings reduce the sound that travels to residents before getting into the building, property, courtyard, yard, or room.

Image 7.62 **An interstate highway that has trees, shrubs, plants, and grasses designed along the sides and center. The greening reduces the sound waves from the vehicles to neighboring communities.**

7.4.6 Distance of Plants to Source of the Sound

(Famulari, 2020)

Similar to the information about air contaminants, there are several factors in determining the distance from the source of the sound to see how far it can spread. The same factors include the current wind, the wind pattern, the structures that may block the air passage (such as walls or buildings), the slope of the land, the current weather, and the frequency and sound wave caused by the source itself. Some additional factors that affect sound include the materials of the structures and frequency of the sound, as well as its volume.

Surfaces that are soft, dense, and have air pockets absorb sound. Foliage and sponges can be used as soundproofing material in buildings. The frequency of the sound itself also affects how far it is carried. Low-frequency sounds such as bass sounds can shake and vibrate much further than higher-frequency sounds such as birds chirping. One other factor that helps determine how much a sound affects you is the context in which you hear the noise. For example, trucks moving in the middle of day with other

Images 7.63–7.68 Green design solutions for equity and inclusion in public community spaces are important for moving forward toward a culture of success for the society of people to be heard, be healthy, and thrive.

bustling traffic is less likely to bother you than a single loud truck of the same frequency in the middle of a quiet night. Though the frequency of the sound is the same, the context shift has a greater effect on you.

7.5 Conclusion

Equality and Inclusion in Green Design: Inclusive Approaches for Designing Healthier Work and Living Spaces focuses on seeing different spaces for the greening they can be, as well as teaching people to understand how to see unique ideas toward equity and inclusion in design. Green design is diverse and ever-changing, and it can be used to address some of the inequity and exclusion that is present in private and public spaces.

As there are more modalities and the needs and locations where people now work, making sure that multiple types of spaces are designed for the diversity of people's success is more relevant than ever. This includes designs for spaces that are urban, suburban, rural, towns, offices, commercial spaces, warehouses, plazas, homes, studios, and more.

This book gives readers a way to not only understand greening, but also to understand how to see greening applied to their site specifically toward creating equitable and inclusive design for the strength of their business, health center, university, education center, home, street, neighborhood, community, and region.

Understanding where there are design challenges can lead to even greater change of design solutions. It is in the solutions where equity and inclusion are most effective and can bring people, neighbors, and communities together to raise the standard of health, abilities, sexuality, gender, age, race, language, income, and cultures toward success. Success leads to greater shared growth for people to be seen, heard, celebrated, and given green spaces to be able to thrive and flourish in their lives.

References

Amato, F., Pandolfi, M., Viana, M., Querol, X., Alastuey, A., & Moreno, T. (2009). Spatial and Chemical Patterns of PM10 in Road Dust Deposited in Urban Environment. *Institute of Earth Sciences*, Jaume Almera, Spanish Research Council (CSIC), Barcelona, Spain.

Barnes, J. (2021). Gendered Landscapes. *The Field, American Society of Landscape Architects*. Retrieved 20 December 2023, from https://thefield.asla.org/2021/06/08/gendered-landscapes/

Belis, C. A., Peroni, E., & Thunis, P. (2013). *Source Apportionment of Air Pollution in the Danube Region*. http://iet.jrc.ec.europa.eu

Boomsma, C., & Steg, L. (2014). Feeling Safe in the Dark: Examining the Effect of Entrapment, Lighting Levels, and Gender on Feelings of Safety and Lighting Policy Acceptability. *Environmental and Behaviour*, 46(2), 193–212.

Chapman, R. (n.d.). *Parks and an Equitable Recovery*. Trust for Public Land. Retrieved 16 February 2025, from https://www.tpl.org/parks-and-an-equitable-recovery-parkscore-report#:~:text=Our%20data%20shows%20major%20disparities,in%20park%20investment%20and%20planning

Cohen, D. A., Williamson, S., & Han, B. (2021). Gender Differences in Physical Activity Associated with Urban Neighborhood Parks: Findings from the National Study of Neighborhood Parks. *Women's Health Issues: Official Publication of the Jacobs Institute of Women's Health*, 31(3), 236–244. https://doi.org/10.1016/j.whi.2020.11.007

Etopio, A. L., Devereux, P., & Crowder, M. (2019). Perceived Campus Safety as a Mediator of the Link Between Gender and Mental Health in a National U.S. College Sample. *Women Health*, 59(7), 703–717.

Famulari, S. (2020). *Green Up! Sustainable Design Solutions for Healthier Work and Living Environments*. New York: Productivity Press. https://doi.org/10.4324/978042297434

Famulari, S. (2024). *Ways of Greening. Using Plants and gardens for Healthy Work and Living Surroundings*. New York: Productivity Press. https://doi.org/10.4324/9781003348634

Fernandez, M. (2006, October 29). A Study Links Trucks' Exhaust to Bronx School Children's Asthma. *The New York Times*, Region Section.

Foster, S., Giles-Corti, B., & Knuiman, M. (2014). Does Fear of Crime Discourage Walkers? A Social-Ecological Exploration of Fear as a Deterrent to Walking. *Environmental and Behaviour*, 46(6), 698–717.

Gjording, L. R. (2022). Redlining and Its Impact on New York City. *City Signal, Columbia University School of Public Health*. Real Estate Section. Retrieved 20 December 2023, from https://www.citysignal.com/redlining-and-its-impact-on-new-york-city/

Goodyear, S. (2015). More Women Ride Mass Transit Than Men. Shouldn't Transit Agencies Be Catering to Them? *Bloomberg*. Retrieved 20 December 2023, from https://www.bloomberg.com/news/articles/2015-01-30/more-women-ride-mass-transit-than-men-shouldn-t-transit-agencies-be-catering-to-them

Hornikx, M. (2016). Ten Questions Concerning Computational Urban Acoustics. *Building and Environment*, 106, 409–421.

Hunt, E. (2019). City with a Female Face: How Modern Vienna was Shaped by Women. *The Guardian*. Retrieved 20 December 2023, from https://www.theguardian.com/cities/2019/may/14/city-with-a-female-face-how-modern-vienna-was-shaped-by-women

Karagulian, F., Belis, C. A., Dora, C. F., Pruss-Ustan, A. M., Bonjour, S., Adair-Roahn, H., & Amann, M. (2015). Contributions to Cities' Ambient Particulate Matter (PM): A Systematic Review of Local Source Contributions at Global Level. *Atmospheric Environment*, 120, 475–483.

Loewen, L. J., Steel, G. D., & Suedfeld, P. (1993). Perceived Safety from Crime in the Urban Environment. *Journal of Environmental Psychology*, 13(4), 323–331.

Mittan, K. (2020, October 5). Black and Hispanic People More Likely to Live in High-Risk Flood Zones, Study Finds. *University of Arizona News*. Retrieved 20 December 2023, from https://news.arizona.edu/story/black-and-hispanic-people-more-likely-live-high-risk-flood-zones-study-finds

Nagendra, H., & Sen, A. (2022). Our Cities are Designed by Men, for Men. *The Third Eye and Science the Wire*. Retrieved 20 February 2025, from https://science.thewire.in/society/gender/our-cities-are-designed-for-men-by-men/

Nardone, A., Rudolph, K. E., Rachel Morello-Frosch, R., & Casey, J. A. (2021). Redlines and Greenspace: The Relationship Between Historical Redlining and 2010 Greenspace Across the United States. *Environmental Health Perspectives*, 129(1), 1. https://doi.org/10.1289/EHP7495

New York State Department of Health. (n.d.). Retrieved from https://www.health.ny.gov/environmental/indoors/air/pmq_a.htm

Pool, J. R., Grinspan, D., Trivedi, A., Anderson, J., & Bouye, M. (2020). Green Space: An Underestimated Tool to Create More Equal Cities. *World Resources Institute*. Retrieved 20 December 2023, from https://www.wri.org/insights/green-space-underestimated-tool-create-more-equal-cities

Public Health Columbia University. (2021). *Historically Redlined Neighborhoods are More Likely to Lack Green Space Today: Study*. Public Health, Columbia Mailman School of Public Health. Retrieved 20 December 2023, from https://www.publichealth.columbia.edu/news/historically-redlined-neighborhoods-are-more-likely-lack-green-space-today-study

Rustin, S. (2014). If Women Built Cities, What Would Our Urban Landscape Look Like? *The Guardian*. Retrieved 20 December 2023, from https://www.theguardia.com/cities/2014/dec/05/if-women-built-cities-what-would-our-urban-landscapes-look-like

Sauer, E. (2022). The Gender Gap in Caregiving and Why Women Carry it. *UMKC Word Press*. Retrieved 25 February 2025, from https://info.umkc.edu/womenc/2022/03/09/the-gender-gap-in-caregiving-and-why-women-carry-it/

Scrader, A. (2023). *Study: Minorities More Likely Than White People to Live Behind Subpar Levees*. Retrieved 20 December 2023, from https://www.upi.com/Science_News/2023/09/05/minorities-live-behind-subpar-levees/6131693965349/#:~:text=Those%20figures%20can%20then%20be,in%20the%20Midwest%20and%20Southeast

Takin, M., Cilliers, E. J., & Ghosh, S. (2023). Advancing Flood Resilience: The Nexus Between Flood Risk Management, Green Infrastructure, and Resilience. *Frontiers in Sustainable Cities*, 5, 1186885. https://doi.org/10.3389/frsc.2023.1186885. Retrieved 20 December 2023, from http://www.frontiersin.org/articles/10.3389/frsc.2023.1186885/full

Ward, Jr., K. (21 December 2021). How Black Communities Become "Sacrifice Zones" for Industrial Air Pollution. *ProPublica*. Retrieved 16 February 2025, from https://www.propublica.org/article/how-black-communities-become-sacrifice-zones-for-industrial-air-pollution

Wolf, K. L., & Flora, K. (2010). Mental Health and Function – A Literature Review. In: *Green Cities: Good Health*. College of the Environment, University of Washington. www.greenhealth.washington.edu

Woods, Z. (2016). *Last Woman Standing: Toilet Equality for Men and Women*. Retrieved 25 February 2025, from https://blogs.adb.org/blog/last-woman-standing-toilet-equality-men-and-women

World Health Organization. (2016). *Urban Green Spaces and Health*. Geneva: WHO Regional Office of Europe.

Yang, X., Wang, Z., Zhigang, L., & Tang, Z. (2017). An Assessment of Urban Park Access in Shanghai-Implications for the Social Equity in Urban China. *Landscape and Urban Planning*, 157, 383–393. Retrieved 20 December 2023, from https://doi.org/10.1016/j.landurbplan.2016.08.007

Zippia. (n.d.). *Designer Demographics and Statistics in the US*. Retrieved 25 February 2025, from https://www.zippia.com/designer-jobs/demographics/

Bibliography

Abbit, L. (2015). *Playgrounds for Seniors Coming Your Way*. Senior Planet, AARP. Retrieved 20 December 2023, from https://seniorplanet.org/playgrounds-for-seniors/

Acolin, A., Crowder, K., Decter-Frain, A., Hajat, A., & Hall, M. (2023). Gentrification, Mobility, and Exposure to Contextual Determinants of Health. *Housing Policy Debate,* 33(1), 194–223. Retrieved 20 December 2023, from https://www.tandfonline.com/doi/full/10.1080/10511482.2022.2099937

Ajzurcher. (2024). *A Greener Future for Everyone: Greenspace Inequity in San Francisco* Retrieved 23 April 2024, from https://usfblogs.usfca.edu/sustainability/2024/04/22/a-greener-future-for-everyone-greenspace-inequality-in-san-francisco/

Allen, D. J. (2023). Confronting the Racist Legacy of Urban Highways. The Dirt, Uniting the Built and Natural Environments. *The American Society of Landscape Architects.* Retrieved 20 December 2023, from https://dirt.asla.org/2023/10/02/confronting-the-racist-legacy-of-urban-highways/

Alonso, E. B., Cockx, L., & Swinnen, J. (2018). Culture and Food Security. *Global Food Security*, 17, 113–127. ISSN 2211–9124. https://doi.org/10.1016/j.gfs.2018.02.002

Amato, F., Pandolfi, M., Escrig, A., Querol, X., Alastuey, A., Pey, J., Perez, N., & Hopke, P. K. (2009). Quantifying Road Dust Resuspension in Urban Environment by Multilinear Engine: A Comparison with PMF2. *Atmospheric Environment*, 43(17), 2770–2780.

Amato, F., Pandolfi, M., Viana, M., Querol, X., Alastuey, A., & Moreno, T. (2009). Spatial and Chemical Patterns of PM10 in Road Dust Deposited in Urban Environment. *Institute of Earth Sciences*, Jaume Almera, Spanish Research Council (CSIC), Barcelona, Spain.

American Chemical Society. (2009, July 23). Stop and Smell the Flowers – the Scent Really Can Soothe Stress. *ScienceDaily*. Retrieved 26 April 2023, from www.sciencedaily.com/releases/2009/07/090722110901.htm

American Society of Landscape Architects. (2015). Fayetteville 2030: Food City Scenario. *ASLA 2015 Professional Awards.* Retrieved 20 December 2023, from https://www.asla.org/2015awards/94716.html

American Society of Landscape Architects. (2021). Farm for the City, Philadelphia, Pennsylvania, Viridian Landscape Studio. *ASLA 2021 Professional Awards.* Retrieved 20 December 2023, from https://www.asla.org/2021awards/2693.html

Assad, R., & Jezzini, Y. (2024). Assessing and Predicting Green Gentrification Susceptibility Using an Integrated Machine Learning Approach. *Local Environment the International Journal of Justice and Sustainability.* Retrieved 20 January 2025, from https://doi.org/10.1080//13549839.2024.2353058

Austin Turner, M., & Greene, S. (n.d.). Urban Institute. Retrieved 16 February 2025, from https://www.urban.org/racial-equity-analytics-lab/structural-racism-explainer-collection/causes-and-consequences-separate-and-unequal-neighborhoods

Australian Institute of Landscape Architects, AILA. (2018). Ballum Bumps Regional Play Space. *AILA 2019 National Landscape Architecture Award*. Retrieved 20 December 2023, from https://www.landscapearchitectureprojects.com/projects/2018/ballumbumps

Bainnson, B. (n.d.). Ed Benedict Park and Portland Memory Garden. *American Society of Landscape Architects*. Retrieved 20 December 2023, from https://www.asla.org/portland/site.aspx?id=43611

Balta-Ozkan, N., Watson, T., & Mocca E. (2015). Spatially Uneven Development and Low Carbon Transitions: Insights From Urban and Regional Planning. *Energy Policy*, 85. 500–510. Retrieved 20 December 2023, from https://doi.org/10.1016/j.enpol.2015.05.013

Barnes, J. (2021). Gendered Landscapes. *The Field, American Society of Landscape Architects*. Retrieved 20 December 2023, from https://thefield.asla.org/2021/06/08/gendered-landscapes/

Bayou City Waterkeeper. (2020). Flooding Disproportionately Impacts People of Color. *Bayou Cit Waterkeeper*. Retrieved 20 December 2023, from https://bayoucitywaterkeeper.org/flooding-disproportionately-impacts-people-of-color/#:~:text=One%20year%20after%20Harvey%2C%20more,unable%20to%20reinhabit%20their%20homes

Belis, C. A., Peroni, E., & Thunis, P. (2013). *Source Apportionment of Air Pollution in the Danube Region*. http://iet.jrc.ec.europa.eu

Berger, M. (2019). A Guide to How Gender Neutral Language is Developing Around the World. *The Washington Post*. Retrieved 20 December 2023, from https://www.washingtonpost.com/world/2019/12/15/guide-how-gender-neutral-language-is-developing-around-world/

Biggs, C. (2019, May 10). Rooms with a View (and How Much You'll Pay for Them). *The New York Times*.

Boomsma, C., & Steg, L. (2014). Feeling Safe in the Dark: Examining the Effect of Entrapment, Lighting Levels, and Gender on Feelings of Safety and Lighting Policy Acceptability. *Environmental and Behaviour*, 46(2), 193–212.

Boone, K. (2023, February 15). *Recognition, Reconciliation, Reparation* [Lecture]. Harvard Graduate School of Design. Retrieved from https://www.gsd.harvard.edu/event/kofi-boone-recognition-reconciliation-reparation/

Bosch, A., Robles, E. H., & Pool, J. R. (2021). Improving Water Security Helps Reduce the Gender Gap in Mexico. *Cities 4 Forests*. Retrieved 20 December 2023, from https://cities4forests.com/story/improving-water-security-helps-reduce-the-gender-gap-in-mexico-city/

Buenos Aires. (n.d.). Buenos Aires Strengthens Natural Resources to Protect Low-Income Communities. *Inclusive Climate Action – in Practice*. Retrieved 20 December 2023, from https://cdn.locomotive.works/sites/5ab410c8a2f42204838f797e/content_entry5ab410fb74c4833febe6c81a/5c4205955f26f200194371b5/files/Buenos_Aires.pdf?1547830677

Burdette, A. M., & Needham, B. L. (2012). Neighborhood Environment and Body Mass Index Trajectories From Adolescence to Adulthood. *Journal of Adolescent Health*, 50(1), 30–37.

Cattarall, P., & Azzouz, A. (2021). The Queer City: How to Design More Inclusive Public Space. *The Conversation US*. Retrieved 20 December 2023, from https://theconversation.com/the-queer-city-how-to-design-more-inclusive-public-space-161088

CDC. (n.d.). Retrieved 2 January 2025, from https://www.cdc.gov/about/priorities/why-is-addressing-sdoh-important.html#cdc_generic_section_3-public-health-actions-that-affect-sdoh

Chaney, R., Baer, A., & Tovar, A. L. (2024). Gender-Based Heat Map Images of Campus Walking Settings: A Reflection of Lived Experience. *Violence and Gender*, 11(1), 35–42. Retrieved 1 June 2024, from https://www.liebertpub.com/action/showCitFormats?doi=10.1089%2Fvio.2023.0027

Chapman, R. (n.d.). *Parks and an Equitable Recovery*. Trust for Public Land. Retrieved 16 February 2025, from https://www.tpl.org/parks-and-an-equitable-recovery-parkscore-report#:~:text=Our%20data%20shows%20major%20disparities,in%20park%20investment%20and%20planning

Cheng, T. (2013). Forms, Transitions, and Design Approaches: Women as Creators of Built Landscapes. *Theses for University of Massachusetts Amherst*. Retrieved 20 December 2023, from https://doi.org/10.7275/4490143

Cities 4 Forests. (n.d.). Retrieved 20 December 2023, from https://cities4forests.com/project/jakarta-tree-management-and-parks-regulations/

Clean Air Council. (2022). Avoiding Economic Displacement During Greening in Philadelphia. *Clean Air Council*. Retrieved 20 December 2023, from https://cleanair.org/greengentrification/

Cohen, D. A., Williamson, S., & Han, B. (2021). Gender Differences in Physical Activity Associated with Urban Neighborhood Parks: Findings from the National Study of Neighborhood Parks. *Women's Health Issues: Official Publication of the Jacobs Institute of Women's Health*, 31(3), 236–244. https://doi.org/10.1016/j.whi.2020.11.007

Coral, F. (2020). *Informal Gardens. Social and Political Realities in a Modernist Brazilian Landscape*. Retrieved 20 December 2023, from https://issuu.com/felipecoral7/docs/informal_gardens

Cummings, B. E., & Waring, M. S. (2020). Potted Plants Do Not Improve Indoor Air Quality: A Review and Analysis of Reported VOC Removal Efficiencies. *Journal of Exposure Science and Environmental Epidemiology*, 30(2), 253–261. https://doi.org/10.1038/s41370-019–0175–9

Design Workship. (n.d.). Retrieved 20 December 2023, from https://www.design-workshop.com/projects/fontana-urban-greening.html

Ding, L., & Hwang, J. (2020). Unequal Displacement: Gentrification, Racial Stratification, and Residential Destinations in Philadelphia. *American Journal of Sociology*, 126(2).

Dirtworks Landscape Architecture PC. (n.d.). Retrieved 20 December 2023, from https://dirtworks.us/portfolio/life-enrichment-center/

Dirtworks Landscape Architecture PC. (n.d.). Retrieved 20 December 2023, from https://dirtworks.us/portfolio/sensory-arts-garden-els-center-excellence/

Eden Green Techonology. (n.d.). Retrieved 20 December 2023, from https://www.edengreen.com/our-greenhouses

Els for Autism. (n.d.). Retrieved 20 December 2023, from https://www.elsforautism.org/the-els-center-of-excellence/sensory-arts-garden/

Etopio, A. L., Devereux, P., & Crowder, M. (2019). Perceived Campus Safety as a Mediator of the Link Between Gender and Mental Health in a National U.S. College Sample. *Women Health*, 59(7), 703–717.

Evans, F. (2021, October 20; updated 2023, September 21). How Interstate Highways Gutted Communities- and Reinforced Segregation. *History, A&E Television Network*. Retrieved 20 December 2023, from https://www.history.com/news/interstate-highway-system-infrastructure-construction-segregation

Famulari, S. (2020). *Green Up! Sustainable Design Solutions for Healthier Work and Living Environments*. New York: Productivity Press. https://doi.org/10.4324/978042297434

Famulari, S. (2024). *Ways of Greening. Using Plants and Gardens for Healthy Work and Living Surroundings*. New York: Productivity Press. https://doi.org/10.4324/9781003348634

Feder, S. (2020, December 1). *Stanford Professor's Study Finds Gentrification Disproportionately Affects Minorities*. Standford Report, Standford University Social Sciences. Retrieved 25 February 2025, from https://news.stanford.edu/stories/2020/12/01/gentrification-disproportionately-affects-minorities

Fernandez, M. (2006, October 29). A Study Links Trucks' Exhaust to Bronx School Children's Asthma. *The New York Times*, Region Section.

Fieldstadt, E. (2016). North Carolina Bathroom Law Give Gender-Neutral Clothing New Purpose. *NBC News*. Retrieved 20 December 2023, from https://www.nbcnews.com/news/us-news/north-carolina-bathroom-law-gives-gender-neutral-clothing-logo-new-n571391

First Street Foundation. (n.d.). Retrieved 20 December 2023, from https://riskfactor.com/solutions/flood#personalsolutions

Fooks, M. (2021). Queering Landscape Architecture, Deconstructing Binaries. *Landscape Australia*. Retrieved 20 December 2023, from https://landscapeaustralia.com/articles/queer-ecologies-review/

Foster, S., Giles-Corti, B., & Knuiman, M. (2014a) Does Fear of Crime Discourage Walkers? A Social-Ecological Exploration of Fear as a Deterrent to Walking. *Environmental and Behaviour*, 46(6), 698–717.

Gjording, L. R. (2022). Redlining and Its Impact on New York City. *City Signal, Columbia University School of Public Health*. Real Estate Section. Retrieved 20 December 2023, from https://www.citysignal.com/redlining-and-its-impact-on-new-york-city/

Gluck, M. (2019). Los Angeles Has Become a Hub for a Wave of Queer Street Artists. From commissioned mural to wheatpasted posters, gay art is popping up all over L.A.'s streetscape. *Los Angeles Magazine*. Retrieved 20 December 2023, from https://lamag.com/art/queer-street-art

Goodyear, S. (2015). More Women Ride Mass Transit Than Men. Shouldn't Transit Agencies Be Catering to Them? *Bloomberg*. Retrieved 20 December 2023, from https://www.bloomberg.com/news/articles/2015-01-30/more-women-ride-mass-transit-than-men-shouldn-t-transit-agencies-be-catering-to-them

Gould, K. A., & Lewis, T. (2012). The Environmental Injustice of Green Gentrification: The Case of Brooklyn's Prospect Park. *The World in Brooklyn: Gentrification, Immigration, and Ethnic Politics in a Global City*, 113–146. Retrieved 20 December 2023, from https://www.researchgate.net/publication/285313933_The_environmental_injustice_of_green_gentrification_the_case_of_Brooklyn's_prospect_park

Green, J. (2023). Kofi Boone: Designing a Black Commons. The Dirt, Uniting the Built and Natural Environments. *The American Society of Landscape Architects*. Retrieved 20 December 2023, from https://dirt.asla.org/2023/02/21/kofi-boone-designing-a-black-commons/

Green, J. (2023). Landscape Architect's Perspective on Waters of the U.S. The Dirt, Uniting the Built and Natural Environments. *The American Society of Landscape Architects.* Retrieved 20 December 2023, from https://dirt.asla.org/2023/08/18/landscape-architects-perspectives-on-waters-of-the-u-s/

Green, J. (2023). New Strategies for Preventing Green Gentrification. The Dirt, Uniting the Built and Natural Environments. *The American Society of Landscape Architects.* Retrieved 20 December 2023, from https://dirt.asla.org/2023/01/31/new-strategies-for-preventing-green-gentrification/

Griffis, J. (2023). Finding Your Way: A Guide To Intuitive Wayfinding. *Progressive AE.* Retrieved 20 December 2023, from https://www.progressiveae.com/finding-your-way-a-guide-to-intuitive-wayfinding/

Hale, T. (2022). Plants Could Be the Key to a Better Semester. *Daily Universe.*

Han, K.-T., & Ruan, L.-W. (2019). Effects of Indoor Plants on Self-Reported Perceptions: A Systemic Review. *Sustainability*, 11(16), 4506. http://doi.org/10.3390/su11164506

Hirsch, A. (2021). Food and Climate Justice in the San Joaquin Valley. *Sites, USC.* Retrieved 20 December 2023, from https://sites.usc.edu/landscape-justice-initiative/2021/01/12/a-landscape-ethic-for-the-working-country-histories-practices-futures-of-the-san-joaquin-valley/

Holmes, D. (2020). *Thammasat University – the Largest Urban Rooftop Farm in Asia.* Retrieved 1 December 2024, from https://worldlandscapearchitect.com/thammasat-university-the-largest-urban-rooftop-farm-in-asia/?v=b870c45f9584

Holmes, D. (2021). Designing the Bigger Time: Tools for a Gender Just Workplace. *WxLA, World Landscape Architecture.* Retrieved 20 December 2023, from https://worldlandscapearchitect.com/designing-the-bigger-time-tools-for-a-gender-just-workplace/?v=7516fd43adaa

Hood Design Studio. (n.d.). Lift Ev'ry Voice and Sing Park. Retrieved 20 December 2023, from https://www.hooddesignstudio.com/liftevryvoiceandsingpark

Hornikx, M. (2016). Ten Questions Concerning Computational Urban Acoustics. *Building and Environment*, 106, 409–421.

Hou, J. (2021). Landscape Architects Can Become Changemakers, But First We Need to Change How They Are Taught. The Dirt, Uniting the Built and Natural Environments. *The American Society of Landscape Architects.* Retrieved 20 December 2023, from https://dirt.asla.org/2021/04/19/landscape-architects-can-become-changemakers-but-first-we-need-to-change-how-they-are-taught/

Hunt, E. (2019). City with a Female Face: How Modern Vienna was Shaped by Women. *The Guardian.* Retrieved 20 December 2023, from https://www.theguardian.com/cities/2019/may/14/city-with-a-female-face-how-modern-vienna-was-shaped-by-women

Imrie, R. (2012). Universalism, Universal Design, and Equitable Access to the Built Environment. *Disability & Rehabilitation*, 34(10), 873–882. https://doi.org/10.3109/09638288.2011.624250

Karagulian, F., Belis, C. A., Dora, C. F., Pruss-Ustan, A. M., Bonjour, S., Adair-Roahn, H., & Amann, M. (2015). Contributions to Cities' Ambient Particulate Matter (PM): A Systematic Review of Local Source Contributions at Global Level. *Atmospheric Environment*, 120, 475–483.

Kennedy J. (2023). Power to Pursue Happiness. *Daedalus*, 152(1), 52–57.

Kratz, J. (2024). The Little Known History of DEI and Why it's Critical to its Survival. *Forbes.* Retrieved 20 February 2025, from https://www.forbes.com/sites/juliekratz/2024/12/29/history-of-dei-why-it-matters-for-the-future/

Larson NI, Wall MM, Story MT, et al. (2013)Home/family, peer, school, and neighborhood correlates of obesity in adolescents. *Obesity*, 21(9), 1858–1869.

Loewen, L. J., Steel, G. D., & Suedfeld, P. (1993). Perceived Safety From Crime in the Urban Environment. *Journal of Environmental Psychology*, 13(4), 323–331.

Loukaitou-Sideris, A. (2009). *How to Ease Women's Fear of Transportation Environments: Case Studies and Best Practices.* Retrieved 25 February 2025, from https://transweb.sjsu.edu/research/How-Ease-Womens-Fear-Transportation-Environments-Case-Studies-and-Best-Practices

Maarten Hornikx, M. (2016). Ten Questions Concerning Computational Urban Acoustics. *Building and Environment*, 106, 409–421.

Marston, T. J. (2019). Two Women [Re]Making Wikipedia History. *The Field, The American Sociate of Landscape Architects.* Retrieved 20 December 2023, from https://thefield.asla.org/2019/12/05/two-women-remaking-wikipedia-history/

Mittan, K. (2020, October 5). Black and Hispanic People More Likely to Live in High-Risk Flood Zones, Study Finds. *University of Arizona News.* Retrieved 20 December 2023, from https://news.arizona.edu/story/black-and-hispanic-people-more-likely-live-high-risk-flood-zones-study-finds

NAACP. (n.d.). Retrieved 20 December 2023, from https://naacp.org/resources/criminal-justice-fact-sheet#:~:text=Anything%20you%20say%20can%20and,cause%20than%20a%20white%20person

Nagendra, H., & Sen, A. (2022). Our Cities are Designed by Men, For Men. *The Third Eye and Science the Wire.* Retrieved 20 February 2025, from https://science.thewire.in/society/gender/our-cities-are-designed-for-men-by-men/

Nardone, A., Rudolph, K. E., Rachel Morello-Frosch, R., & Casey, J. A. (2021). Redlines and Greenspace: The Relationship Between Historical Redlining and 2010 Greenspace Across the United States. *Environmental Health Perspectives*, 129(1), 1. https://doi.org/10.1289/EHP7495

Nateras, M. M. (2023). Cultivating Climate, Food and Land Justice. *American Friends Service Committee, AFSC.* Retrieved 20 December 2023, from https://afsc.org/news/cultivating-climate-food-and-land-justice

National Center for Education Statistics. (2024). *Concentration of Public School Students Eligible for Free or Reduced-Price Lunch. Condition of Education.* U.S. Department of Education, Institute of Education Sciences. Retrieved 16 February 2025, from https://nces.ed.gov/programs/coe/indicator/clb

Naturally Resilient Communities. (n.d.). Retrieved 20 December 2023, from https://nrcsolutions.org/

New York State Department of Health. (n.d.). Retrieved from https://www.health.ny.gov/environmental/indoors/air/pmq_a.htm

North Coast Rape Crisis Team. (n.d.). Retrieved 25 February 2025, from https://www.humboldt.edu/supporting-survivors/educational-resources/statistics

NRPA. (n.d.). *Parks and Improved Mental Health and Quality of Life.* Retrieved 20 February 2025, from https://www.nrpa.org/our-work/Three-Pillars/health-and-well-being/ParksandHealth/fact-sheets/parks-improved-mental-health-quality-life/

Nuccitelli, D. (2023). The Little-Known Physical and Mental Health Benefits of Urban Trees. *Yale Climate Connections.* Retrieved 12 March 2023, from https://yaleclimateconnections.org/2023/02/the-little-known-physical-and-mental-health-benefits-of-urban-trees/

Odom, J. (2016, November 21). Researchers Dig Deeper to Discover Healing Power of Green Spaces. *Total Landscape Care, Randall-Reilly.* Retrieved 20 December 2023, from https://www.totallandscapecare.com/business/article/15040235/the-green-road-project-helps-those-struggling-with-ptsd

Park, Y., & Garcia, M. (2020). Pedestrian Safety Perception and Urban Street Settings. *International Journal of Sustainable Transportation,* 14(11), 860–871.

Peachey, A. A., & Baller, S. L. (2015). Perceived Built Environment Characteristics of On-Campus and Off-Campus Neighborhoods Associated with Physical Activity of College Students. *Journal of American College Health*, 63(5), 337–342.

Poobalan, A. S., et al. (2012). Physical Activity Attitudes, Intentions and Behaviour Among 18–25 Year Olds: A Mixed Method Study. *BMC Public Health*, 12(1), 640.

Pool, J. R., Grinspan, D., Trivedi, A., Anderson, J., & Bouye, M. (2020). Green Space: An Underestimated Tool to Create More Equal Cities. *World Resources Institute.* Retrieved 20 December 2023, from https://www.wri.org/insights/green-space-underestimated-tool-create-more-equal-cities

Pool, T. (2022). *On Plant-Associated Microbiota in the Indoors.* Green Architecture, Healthy Plants, Indoor Green. Retrieved 12 March 2023, from https://moss.amsterdam/2022/12/09/on-the-progressive-effects-of-plant-associated-microbiota-on-the-indoor-environment-2/

Public Health Columbia University. (2021). *Historically Redlined Neighborhoods are More Likely to Lack Green Space Today: Study.* Public Health, Columbia Mailman School of Public Health. Retrieved 20 December 2023, from https://www.publichealth.columbia.edu/news/historically-redlined-neighborhoods-are-more-likely-lack-green-space-today-study

Pyzyk, K. (2018). Creative Crosswalks: Street Art Meets Safety Enhancement. Bright Colors and Unique Designs in Crosswalks Can Create a Sense of Community While Keeping Pedestrians Safer and Drawing Drivers' Attention to Them. *Smart Cities Dive.* Retrieved 20 December 2023, from https://www.smartcitiesdive.com/news/creative-crosswalks-street-art-meets-safety-enhancement/526474/

Queer Botany. (n.d.). Retrieved 20 December 2023, from https://www.queerbotany.com/

Quiiky Magazine. (2022). LGBTA+ Street Art: 12 Queer Murals to Admire Around the World. *Quiiky Magazine.* Retrieved 20 December 2023, from https://quiikymagazine.com/en/gay-street-art/

Quintana, M. (2016). Changing Grid: Exploring the Impact of the High Line. *Street Easy Reads.* Retrieved 20 December 2023, from https://streeteasy.com/blog/changing-grid-high-line/

Raanaas, R. K., Horgen Evensen, K., Rich, D., Siostrom, G., & Patil, G. (2011). Benefits of Indoor Plants on Capacity in an Office Setting. *Journal of Environmental Psychology*, 31(1), 99–105.

Ramos Pacheco, M. (2025). Dallas Unveils Plans for First Five Greenspaces to Be Built in Areas with No Nearby Parks, Trust for Public Land Partners with the City of Dallas to Create New Green Space on Unused City-Owned Land. *Dallas News.* Retrieved 25 February 2025, from https://www.dallasnews.com/news/environment/2025/02/25/dallas-unveils-plans-for-first-five-greenspaces-to-be-built-in-areas-with-no-nearby-parks/

Rethinking the Future. (n.d.). Retrieved 20 December 2023, from https://www.re-thinkingthefuture.com/rtf-fresh-perspectives/a10986-queering-urban-landscapes-exploring-lgbtq-inclusive-architecture/

Rierson, S. L. (1994). Race and Gender Discrimination: A Historical Case for Equal Treatment Under the Fourteenth Amendment. *Duke Journal of Gender Law and Policy*, 1, 89–117.

Rustin, S. (2014). If Women Built Cities, What Would Our Urban Landscape Look Like? *The Guardian.* Retrieved 20 December 2023, from https://www.theguardia.com/cities/2014/dec/05/if-women-built-cities-what-would-our-urban-landscapes-look-like

Sasaki Associates. (n.d.). Mississippi River Flood Resiliency Plan. Creating a More Flood Resilient, Accessible, and Beautiful Riverfront for Davenport, Iowa. *Sasaki Associates.* Retrieved 20 December 2023, from https://www.sasaki.com/projects/mississippi-river-flood-resiliency-plan/

Sasaki Associates. (n.d.). NYC Community Park Initiative. Accessible and Sustainable Parks for Underserved Communities. *Sasaki Associates.* Retrieved 20 December 2023, from https://www.sasaki.com/projects/nyc-community-parks-initiative/

Sauer, E. (2022). The Gender Gap in Caregiving and Why Women Carry it. *UMKC Word Press.* Retrieved 25 February 2025, from https://info.umkc.edu/womenc/2022/03/09/the-gender-gap-in-caregiving-and-why-women-carry-it/

SCAPE Landscape Architecture DPC. (2007). Retrieved 20 December 2023, from https://www.scapestudio.com/projects/corktown-framework-plan/

SCAPE Landscape Architecture DPC. (2007). Retrieved 20 December 2023, from https://www.scapestudio.com/projects/2023-coastal-master-plan/

SCAPE Landscape Architecture DPC. (2007). Retrieved 20 December 2023, from https://www.scapestudio.com/projects/resilient-jacksonville/

Schmidt, C. (2022). Tapestried Landscape: The Queer Influence of Roberto Burle Marx on Elizabeth Bishop's Brazil. *Modernism/modernity, John Hopkins University Press*, 6, 3. Retrieved 20 December 2023, from https://modernismmodernity.org/articles/schmidt-tapestried-landscape-queer-influence-roberto-burle-marx-elizabeth-bishops-brazil

Scrader, A. (2023). Study: Minorities More Likely Than White People to Live Behind Subpar Levees. Retrieved 20 December 2023, from https://www.upi.com/Science_News/2023/09/05/minorities-live-behind-subpar-levees/613169396 5349/#:~:text=Those%20figures%20can%20then%20be,in%20the%20Midwest%20 and%20Southeast

Segura, C. (2024). What are Air Purifying Paints (and How Do They Work). Retrieved 25 February 2025, from https://www.mychemicalfreehouse.net/2024/05/what-are-air-purifying-paints-and-how-do-they-work.html

Solsten, E., McClave, D. (1994). *Austria: a country study.* Library of Congress Cataloging-in-Publication Data, ISSN 1057–5294.

Steinfeld, E., & Maisel, J. (2012). *Universal Design: Creating Inclusive Environments.* Wiley.

Stodolska, M., et al. (2013). "I Was Born in the Hood": Fear of Crime, Outdoor Recreation and Physical Activity Among Mexican-American Urban Adolescents. *Leisure Sciences*, 35(1), 1–15.

Stoner, A. E. (2023). The Very Personal Impact of Community Green Spaces. The Dirt, Uniting the Built and Natural Environments. *The American Society of Landscape Architects.* Retrieved 20 December, from https://dirt.asla.org/2023/11/06/the-very-personal-impact-of-community-green-spaces/

SWA Group. (n.d.). Retrieved 20 December 2023, from https://www.swagroup.com/projects/the-memorial-at-harvey-milk-plaza/

Swanson, E. (2023). Gentrification Leads to Worsened Air Quality, Study Shows. *Henry Ford Health.* Retrieved 20 December 2023, from https://www.henryford.com/blog/2023/05/gentrification-leads-to-worsened-air-quality-study-shows#:~:text=A%20variety%20of%20factors%20contribute,leads%20to%20worsened%20air%20quality

Syracuse City Hall. (n.d.). Retrieved 20 December 2023, from https://www.syr.gov/Departments/Engineering/Floodplains#section-1

Syracuse City Hall. (n.d.). Retrieved 20 December 2023, from https://www.syr.gov/Departments/Engineering/Floodplains#section-5

Takin, M., Cilliers, E. J., & Ghosh, S. (2023). Advancing Flood Resilience: The Nexus Between Flood Risk Management, Green Infrastructure, and Resilience. *Frontiers in Sustainable Cities*, 5, 1186885. https://doi.org/10.3389/frsc.2023.1186885. Retrieved 20 December 2023, from http://www.frontiersin.org/articles/10.3389/frsc.2023.1186885/full

Tauberer, J. (2020). What Three Weeks in Japan Taught Me About Public Transit Signage. *Medium*. Retrieved 20 December, from https://joshuatauberer.medium.com/what-three-weeks-in-japan-taught-me-about-public-transit-signage-77bfc2d6e824

Telama, R, et al. (2005). Physical Activity From Childhood to Adulthood: A 21-Year Tracking Study. *American Journal of Preventive Medicine,* 28(3), 267–273.

The Field. (2016). Sensory Gardens. *American Society of Landscape Architects.* Retrieved 20 December 2023, from https://thefield.asla.org/2016/10/25/sensory-gardens/

The Local Sweden. (2013). Sweden Warms to 'Gender Equal' Snow Ploughing. *The Local Sweden*. Retrieved 20 December 2023, from https://www.thelocal.se/20131211/snow-plowing-should-be-gender-equal-greens

The Organic & Natural Paint Co. (n.d.). Retrieved 25 February, from https://organic-naturalpaint.co.uk/natural-paint/air-purifying-paints/#:~:text=What%20are%20air%20purifying%20paints,lime%20curing%20in%20the%20air

Thompson, M. (2023). Taking Up Space in Nature is Liberating. 'We Look Like This,' Say L.A.'s Queer Plant Lovers. *Los Angeles Times.* Retrieved 20 December 2023, from https://www.latimes.com/lifestyle/story/2023-06-08/taking-up-space-in-nature-is-liberating-la-queer-plant-lovers

Trivedi, A., Elwell, N., & Walker, S. (2017). Addressing Women's Burdens from Water Risks Could Help Reduce World Hunger. *World Resources Institute*. Retrieved 20 December 2023, from https://www.wri.org/insights/addressing-womens-burdens-water-risks-could-help-reduce-world-hunger

Ulrich, R. S., Simons, R. F., Losito, B. D., Fiorito, L., Miles, M. A., & Zelson, M. (1991). Stress Recovery During Exposure to Natural and Urban Environments. *Journal of Environmental Psychology*, 11(3), 201–230. https://www.sciencedirect.com/science/article/pii/S0272494405801847

United States Department of Justice. (2020). *ADA Standards for Accessible Design*. https://www.ada.gov/regs2010/2010ADAStandards/2010ADAstandards.htm

University Gardens. (2025). *University of Warsaw.* Retrieved 25 February 2025, from https://en.uw.edu.pl/about-university/must-see-at-the-university/university-gardens/

U.S. Commission on Civil Right. (n.d.). *Not in My Backyard: Executive Order 12,898 and Title VI as Tools for Achieving Environmental Justice.* Retrieved 16 February 2025, from https://www.usccr.gov/files/pubs/envjust/ch2.htm

Vahedifard, F., Azhar, M., & Brown, D. C. (2023). Overrepresentation of Historically Underserved and Socially Vulnerable Communities Behind Levees in the United States. *Earth's Future*, 11, e2023EF003619. https://doi.org/10.1029/2023EF003619

VELA Project. (n.d.). Retrieved 20 December 2023, from http://www.velaproject.org/vision.html

Ward, Jr., K. (2021, December 21). How Black Communities Become "Sacrifice Zones" for Industrial Air Pollution. *ProPublica.* Retrieved 16 February 2025, from https://www.propublica.org/article/how-black-communities-become-sacrifice-zones-for-industrial-air-pollution

Webb, K., et al. (2024, July 31). Neighborhood Resources Associated with Psychological Trajectories and Neural Reactivity to Reward After Trauma. *AMA Psychiatry*, 81(11), 1090–1100. https://doi.org/10.1001/jamapsychiatry.2024.2148

Wells, N., & Evans, G. (2003). A Room with a View Helps Rural Children Deal with Life's Stresses. *Environment and Behavior*, 35(3), 311–330.

Wihanesta, R., Sulaeman, D., Pradana, D., & Pribadi, Y. S. (2021). The Reasons for Jakarta's Frequent Flooding and How Nature-based Solutions (Nbs) Can Help Reduce the Risk. *World Resources Institute*. Retrieved 20 December 2023, from https://wri-indonesia.org/en/insights/reasons-jakartas-frequent-flooding-and-how-nature-based-solutions-nbs-can-help-reduce-risk

Wolf, K. L., & Flora, K. (2010). Mental Health and Function – A Literature Review. In: *Green Cities: Good Health*. College of the Environment, University of Washington. www.greenhealth.washington.edu

Wolverton, B. C., & Wolverton, J. D. (1993). Plants and soil Microorganisms: Removal of Formaldehyde, Xylene and Ammonia From Indoor Air Environment. *Journal of Mississippi Academy of Sciences*, 38(2), 11–15.

Wolverton, B. C., et al. (1989). *A Study of Interior Landscape Plants for Indoor Air Pollution Abatement: An Interim Report*. Mississippi: NASA.

World Health Organization. (2010). *WHO Guidelines for Indoor Air Quality: Selected Pollutants*. Geneva: WHO Regional Office of Europe.

World Health Organization. (2016). *Urban Green Spaces and Health*. Geneva: WHO Regional Office of Europe.

WxLA. (n.d.). Retrieved 20 December 2023, from https://www.wxlandscape.org/about

Yang, X., Wang, Z., Zhigang, L., & Tang, Z. (2017). An Assessment of Urban Park Access in Shanghai-Implications for the Social Equity in Urban China. *Landscape and Urban Planning*, 157, 383–393. Retrieved 20 December 2023, from https://doi.org/10.1016/j.landurbplan.2016.08.007

Zdanowicz, C., & Yan, H. (2022, September 18). US Public Schools Get a D+ for Poor Conditions, and Experts Say Problems are Getting Worse. Here's What Kids are Facing. *CNN*. Retrieved 16 February 2025, from https://www.cnn.com/2022/09/18/us/school-conditions-2022/index.html

Zippia. (n.d.). *Designer Demographics and Statistics in the US*. Retrieved 25 February 2025, from https://www.zippia.com/designer-jobs/demographics/

Image Credits

Author Page

a.1	Michael Everett

Introduction

0.1	Stevie Famulari, Gds	
0.2	Shutterstock	Ruben M Ramos
0.3	Shutterstock	Videoanita
0.4	Stevie Famulari, Gds	

Chapter 1

1.1	Shutterstock	The Studio
1.2	Shutterstock	AVAlex/Shutterstock
1.3	Shutterstock	Fabrizio Misson
1.4	Shutterstock	AnnGaysorn
1.5	Yurika Ban	
1.6	Shutterstock	as-artmedia
1.7	Shutterstock	The Studio
1.8	Shutterstock	Mark D Bailey
1.9	Shutterstock	Loch Earn
1.10	Shutterstock	Fabrizio Misson
1.11	Shutterstock	AVAlex/Shutterstock
1.12	Shutterstock	iredman
1.13	Shutterstock	AnnGaysorn
1.14	Yurika Ban	
1.15	Shutterstock	anju901

1.16	Shutterstock	as-artmedia
1.17	Stevie Famulari, Gds	
1.18	Stevie Famulari, Gds	
1.19	Stevie Famulari, Gds	
1.20	Stevie Famulari, Gds	
1.21	Stevie Famulari, Gds	

Chapter 2

2.1	Stevie Famulari, Gds	
2.2	Stevie Famulari, Gds	
2.3	Stevie Famulari, Gds	
2.4	Stevie Famulari, Gds	
2.5	Stevie Famulari, Gds	
2.6	Stevie Famulari, Gds	
2.7	Shutterstock	Dziurek
2.8	Shutterstock	RossHelen
2.9	Shutterstock	Bangprikphoto
2.10	Shutterstock	Bangprikphoto
2.11	Stevie Famulari, Gds	
2.12	Stevie Famulari, Gds	
2.13	Shutterstock	Sharad Raval
2.14	Shutterstock	Richard Juilliart
2.15	Shutterstock	hikrcn
2.16	Shutterstock	hikrcn
2.17	Shutterstock	Sadik Gulec
2.18	Shutterstock	Vic Hinterlang
2.19	Shutterstock	chiakto
2.20	Shutterstock	Riccardo Mayer
2.21	Stevie Famulari, Gds	
2.22	Stevie Famulari, Gds	
2.23	Pennsylvania Horticulture Society	
2.24	Pennsylvania Horticulture Society	
2.25	Pennsylvania Horticulture Society	
2.26	Pennsylvania Horticulture Society	
2.27	Stevie Famulari, Gds	
2.28	Stevie Famulari, Gds	
2.29	Shutterstock	Lano Lan

2.30	Shutterstock	Ajith Narayan
2.31	Shutterstock	Nikita M production
2.32	Shutterstock	Scharfsinn
2.33	Shutterstock	huang yi fei
2.34	Stevie Famulari, Gds	

Chapter 3

3.1	Stevie Famulari, Gds	
3.2	Stevie Famulari, Gds	
3.3	Stevie Famulari, Gds	
3.4	Stevie Famulari, Gds	
3.5	Stevie Famulari, Gds	
3.6	Stevie Famulari, Gds	
3.7	Stevie Famulari, Gds	
3.8	Shutterstock	Anton Watman
3.9	Shutterstock	WeStudio
3.10	Shutterstock	akslocum
3.11	Shutterstock	Marc Pagani Photography
3.12	Shutterstock	Pattie Steib
3.13	Shutterstock	EchoFree
3.14	Shutterstock	Syariff Hidayatullah
3.15	Shutterstock	Ron Adar
3.16	Stevie Famulari, Gds	
3.17	Stevie Famulari, Gds	
3.18	Shutterstock	dani daniar
3.19	Shutterstock	Creativa Images
3.20	Shutterstock	dani daniar
3.21	Shutterstock	Sudarshan Jha
3.22	Shutterstock	dani daniar
3.23	Shutterstock	Javaistan
3.24	Shutterstock	Beekeepx
3.25	Stevie Famulari, Gds	
3.26	Stevie Famulari, Gds	
3.27	City of Davenport	
3.28	Courtesy of Sasaki	
3.29	Courtesy of Sasaki	

3.30	Courtesy of Sasaki	
3.31	Courtesy of Sasaki	
3.32	Courtesy of Sasaki	
3.33	Courtesy of Sasaki	
3.34	Stevie Famulari, Gds	
3.35	Stevie Famulari, Gds	
3.36	Shutterstock	Leonid Andrukhov
3.37	Shutterstock	Beekeepx
3.38	Shutterstock	Volha Ushanava
3.39	Shutterstock	Fadhila Hasnah AW
3.40	Stevie Famulari, Gds	
3.41	Stevie Famulari, Gds	
3.42	Stevie Famulari, Gds	

Chapter 4

4.1	Stevie Famulari, Gds	
4.2	Stevie Famulari, Gds	
4.3	Stevie Famulari, Gds	
4.4	Stevie Famulari, Gds	
4.5	Stevie Famulari, Gds	
4.6	Stevie Famulari, Gds	
4.7	Stevie Famulari, Gds	
4.8	Nature Sacred	
4.9	Nature Sacred	
4.10	Nature Sacred	
4.11	Nature Sacred	
4.12	Nature Sacred	
4.13	Shutterstock	Barbara Strelar
4.14	Stevie Famulari, Gds	
4.15	Stevie Famulari, Gds	
4.16	Shutterstock	Wulandari Wulandari
4.17	Shutterstock	R.M. Nunes
4.18	Shutterstock	Coleman photographer
4.19	Shutterstock	MC MediaStudio
4.20	Shutterstock	Videoanita
4.21	Shutterstock	John_T
4.22	Stevie Famulari, Gds	

4.23	Stevie Famulari, Gds	
4.24	Shutterstock	Wynn Wygal
4.25	Shutterstock	TheOldhiro
4.26	Shutterstock	John_T
4.27	Shutterstock	Summit Art Creations
4.28	Shutterstock	Creativa Images
4.29	Shutterstock	Weiming Xie
4.30	Stevie Famulari, Gds	
4.31	Stevie Famulari, Gds	
4.32	Shutterstock	trekandshoot
4.33	Shutterstock	Aerial-motion
4.34	Shutterstock	debra millet
4.35	Shutterstock	Trong Nguyen
4.36	Stevie Famulari, Gds	
4.37	Stevie Famulari, Gds	
4.38	Shutterstock	AaronChenPS2
4.39	Shutterstock	Chiu CHING-I
4.40	Shutterstock	elwynn
4.41	Shutterstock	Hi oe
4.42	Shutterstock	Igor Corovic
4.43	Stevie Famulari, Gds	

Chapter 5

5.1	Stevie Famulari, Gds	
5.2	Stevie Famulari, Gds	
5.3	Stevie Famulari, Gds	
5.4	Stevie Famulari, Gds	
5.5	Stevie Famulari, Gds	
5.6	Stevie Famulari, Gds	
5.7	dsl.richmond	
5.8	dsl.richmond	
5.9	Yurika Ban	
5.10	Shutterstock	Roschetzky Photography
5.11	Stevie Famulari, Gds	
5.12	Stevie Famulari, Gds	
5.13	Google Earth	Mollie Phillips
5.14	Google Earth	Jeff Anders

5.15	Google Earth	Christopher Mandeville
5.16	Google Earth	Theodore Hillman
5.17	Stevie Famulari, Gds	
5.18	Stevie Famulari, Gds	
5.19	Shutterstock	Zack Frank
5.20	Shutterstock	matteo_it
5.21	Shutterstock	arthurgphotography
5.22	Shutterstock	Sandra Foyt
5.23	Stevie Famulari, Gds	
5.24	Stevie Famulari, Gds	
5.25	Stevie Famulari, Gds	

Chapter 6

6.1	Stevie Famulari, Gds	
6.2	Stevie Famulari, Gds	
6.3	Stevie Famulari, Gds	
6.4	Stevie Famulari, Gds	
6.5	Stevie Famulari, Gds	
6.6	Stevie Famulari, Gds	
6.7	Stevie Famulari, Gds	
6.8	Shutterstock	Lucas Alvarez Canga
6.9	Shutterstock	Elena Rostunova
6.10	Shutterstock	Rolf G Wackenberg
6.11	Shutterstock	Ruben M Ramos
6.12	Stevie Famulari, Gds	
6.13	Stevie Famulari, Gds	
6.14	Shutterstock	Oleg Elkov
6.15	Shutterstock	Bruno Passigatti
6.16	Shutterstock	Photo Art Wall Decoration
6.17	Shutterstock	nnattalli
6.18	Shutterstock	edusma7256
6.19	Shutterstock	Leeyakorn06
6.20	Shutterstock	DeeillustratorID
6.21	Stevie Famulari, Gds	
6.22	Stevie Famulari, Gds	
6.23	Shutterstock	TK Kurikawa

6.24	Shutterstock	Alexandra Lande
6.25	Shutterstock	Boris-B
6.26	Shutterstock	Mark D Bailey
6.27	Stevie Famulari, Gds	
6.28	Stevie Famulari, Gds	
6.29	Shutterstock	Elizabeth Foster
6.30	Shutterstock	Anikin Dmitrii
6.31	Shutterstock	Alexander Osenniy
6.32	Shutterstock	Rolf G Wackenberg
6.33	Google Earth	Asukal
6.34	Google Earth	Jorg Meier
6.35	Shutterstock	maodoltee
6.36	Yurika Ban	
6.37	Yurika Ban	
6.38	Shutterstock	ChamanExperience
6.39	Shutterstock	Loch Earn
6.40	Stevie Famulari, Gds	
6.41	Stevie Famulari, Gds	
6.42	Shutterstock	Jordan Tan
6.43	Shutterstock	ChameleonsEye
6.44	Shutterstock	KenSoftTH
6.45	Shutterstock	zhu difeng
6.46	Shutterstock	KenSoftTH
6.47	Shutterstock	RPA Studio
6.48	Shutterstock	sofirinaja
6.49	Shutterstock	Wirestock Creators
6.50	Shutterstock	Andrii Iemelianenko
6.51	Shutterstock	De Visu
6.52	Shutterstock	Rawpixel.com
6.53	Shutterstock	Andrii Zastrozhnov
6.54	Shutterstock	Lopolo
6.55	Shutterstock	Lordn
6.56	Shutterstock	Olya_Beli_Art
6.57	Shutterstock	eamesBot
6.58	Stevie Famulari, Gds	
6.59	Stevie Famulari, Gds	
6.60	Stevie Famulari, Gds	

Chapter 7

7.1	Shutterstock	photo.ua
7.2	Shutterstock	Artist Nadia Russ
7.3	Shutterstock	Holli
7.4	Shutterstock	wavebreakmedia
7.5	Shutterstock	dkdkdkdk
7.6	Shutterstock	Kmarsh
7.7	Shutterstock	Alizada Studios
7.8	Shutterstock	Carey Davis
7.9	Shutterstock	Boy Anthony
7.10	Shutterstock	Toto Santiko Budi
7.11	Shutterstock	photo.ua
7.12	Shutterstock	Trong Nguyen
7.13	Yurika Ban	
7.14	Yurika Ban	
7.15	Yurika Ban	
7.16	Shutterstock	Aria sandi hasim
7.17	Shutterstock	Elizabeth Foster
7.18	Shutterstock	Jenari
7.19	Shutterstock	Artist Nadia Russ
7.20	Shutterstock	Holli
7.21	Shutterstock	Aleksey Matrenin
7.22	Shutterstock	Elizaveta Galitckaia
7.23	Shutterstock	Little Vignettes Photo
7.24	Yurika Ban	
7.25	Shutterstock	Iredman
7.26	Shutterstock	wavebreakmedia
7.27	dsl.richmond	
7.28	Shutterstock	dkdkdkdk
7.29	Shutterstock	Kmarsh
7.30	Shutterstock	MaksEvs
7.31	Shutterstock	kinder my
7.32	Shutterstock	Jose Luis Stephens
7.33	Shutterstock	Jevanto Productions
7.34	Shutterstock	Storm Is Me
7.35	Shutterstock	AlessandraJ
7.36	Shutterstock	travelview
7.37	Shutterstock	RonaldL

7.38	Shutterstock	Brandon Olafsson
7.39	Shutterstock	Kit Leong
7.40	Shutterstock	Alizada Studios
7.41	Shutterstock	anju901
7.42	Shutterstock	New Africa
7.43	Shutterstock	Thx4Stock team
7.44	Shutterstock	DesignPrax
7.45	Shutterstock	Kohyao
7.46	Shutterstock	Boy Anthony
7.47	Shutterstock	Neuiiza
7.48	Shutterstock	Toa55
7.49	Shutterstock	Kate Scott
7.50	Shutterstock	Simone Hogan
7.51	Shutterstock	Papin Lab
7.52	Shutterstock	All for you friend
7.53	Shutterstock	AY Amazefoto
7.54	Shutterstock	sergey lavrishchev
7.55	Shutterstock	Cheng Wei
7.56	Shutterstock	Den Tramper
7.57	Shutterstock	Carey Davis
7.58	Shutterstock	Francesco Scatena
7.59	Shutterstock	Sergii Rudiuk
7.60	Shutterstock	Fouad A. Saad
7.61	Shutterstock	Fouad A. Saad
7.62	Shutterstock	Cars and Travels
7.63	Stevie Famulari, Gds	
7.64	Stevie Famulari, Gds	
7.65	Stevie Famulari, Gds	
7.66	Stevie Famulari, Gds	
7.67	Stevie Famulari, Gds	
7.68	Stevie Famulari, Gds	

Index

Note: Page numbers in *italics* indicate a figure and page numbers in **bold** indicate a table on the corresponding page.

For Product Safety Concerns and Information please contact our EU
representative GPSR@taylorandfrancis.com
Taylor & Francis Verlag GmbH, Kaufingerstraße 24, 80331 München, Germany

www.ingramcontent.com/pod-product-compliance
Ingram Content Group UK Ltd.
Pitfield, Milton Keynes, MK11 3LW, UK
UKHW050059200926
475702UK00019B/2916
9781032757063